VALUES-BASED SAFETY PROCESS

SECOND EDITION

VALUE-BASED SAFETY PROCESS

Improving Your Safety Culture With Behavior-Based Safety

SECOND EDITION

Terry E. McSween, Ph.D.
President and CEO
Quality Safety Edge

A John Wiley & Sons, Inc., Publication

Copyright © 2003 by John Wiley & Sons, Inc. All rights reserved.

Published by John Wiley & Sons, Inc., Hoboken, New Jersey.
Published simultaneously in Canada.

No part of this publication may be reproduced, stored in a retrieval system, or transmitted in any form or by any means, electronic, mechanical, photocopying, recording, scanning, or otherwise, except as permitted under Section 107 or 108 of the 1976 United States Copyright Act, without either the prior written permission of the Publisher, or authorization through payment of the appropriate per-copy fee to the Copyright Clearance Center, Inc., 222 Rosewood Drive, Danvers, MA 01923, 978-750-8400, fax 978-750-4470, or on the web at www.copyright.com. Requests to the Publisher for permission should be addressed to the Permissions Department, John Wiley & Sons, Inc., 111 River Street, Hoboken, NJ 07030, (201) 748-6011, fax (201) 748-6008, e-mail: permreq@wiley.com.

Limit of Liability/Disclaimer of Warranty: While the publisher and author have used their best efforts in preparing this book, they make no representations or warranties with respect to the accuracy or completeness of the contents of this book and specifically disclaim any implied warranties of merchantability or fitness for a particular purpose. No warranty may be created or extended by sales representatives or written sales materials. The advice and strategies contained herein may not be suitable for your situation. You should consult with a professional where appropriate. Neither the publisher nor author shall be liable for any loss of profit or any other commercial damages, including but not limited to special, incidental, consequential, or other damages.

For general information on our other products and services please contact our Customer Care Department within the U.S. at 877-762-2974, outside the U.S. at 317-572-3993 or fax 317-572-4002.

Wiley also publishes its books in a variety of electronic formats. Some content that appears in print, however, may not be available in electronic format.

Library of Congress Cataloging-in-Publication Data is available.

ISBN 0-471-22049-3

Printed in the United States of America

10 9 8 7 6 5 4 3

CONTENTS

Foreword by E. Scott Geller — xi

Preface — xiii

Acknowledgments — xv

CHAPTER 1 Safety Basics — 1

1.1 Traditional Safety Programs   1
1.2 Du Pont's Success   4
1.3 Our Findings   6
1.4 The Safety Triangle   8
1.5 Complacency   8
1.6 Safety as a Team Process   9
1.7 Common Problems with Safety Efforts   11
1.8 Problems with Punishment   13
1.9 Appropriate Use of Punishment   16
1.10 Components of a Proven Safety Process   17

CHAPTER 2 The Vision — 19

2.1 A Typical Week   19

CHAPTER 3 Value-Based Behavioral Safety Process — 21

3.1 A Question of Balance   21
3.2 Ensure a Clear Mission or Vision Statement   24
3.3 Types of Mission and Vision Statements   25
3.4 What Is a Value?   26
3.5 Why Clarify Values?   27
3.6 Use a Proven Process and Build on Basic Values   27
3.7 Concluding Remarks on Values   28

CHAPTER 4 Behavioral Safety Process — 29

4.1 What's in a Name?   30
4.2 Team-Based Process   30
4.3 Programs Versus Process   32

CHAPTER 5 Safety Assessment 33

5.1 What Is a Safety Assessment? 34
5.2 Why Conduct a Safety Assessment? 34
5.3 Who Should Conduct the Assessment? 35
5.4 Objectives of the Assessment 36
5.5 Outcome of the Assessment 39
5.6 Safety Assessment Process 40

    How Do You Conduct a Safety Assessment? 40
    Step 1: Review Safety Data 41
    Step 2: Conduct Interviews 42
    Step 3: Observe Safety Meetings, Safety Audits, and Safety Practices in Work Areas 46
    Step 4: Analyze Information and Develop and Improvement Plan 48
    Step 5: Make the Final Report and Presentation 49

CHAPTER 6 Management Overview and Initial Workshops 51

6.1 What Is the Management Overview? 52
6.2 What Are the Objectives of the Management Overview? 52
6.3 What Is the Design Team Workshop? 53
6.4 What Are the Objectives of the Design Team Workshop? 53
6.5 What Is the Agenda? 53

CHAPTER 7 Final Design 55

7.1 What Are the Objectives of This Phase? 56
7.2 Design Team Process 57
7.3 Role of the Site Management Team 59
7.4 What Are the Steps in This Phase? 59

CHAPTER 8 Step 1: Establishing Mission, Values, and Milestone Targets 61

8.1 Clarifying Values: A Structured Approach 62
8.2 Step 1: Brainstorm Actions Likely to Impact the Process 63
8.3 Step 2: Pinpoint Those Practices 63
8.4 Step 3: Sort These Practices into "Value" Categories 63
8.5 Step 4: Use Values in Designing Your Safety Process 64
8.6 Step 5: Discuss Values During Kickoff Meetings and Training 65
8.7 Step 6: Use Values as Criteria for Evaluation 66
8.8 Establish a Milestone Schedule 66

CHAPTER 9 Step 2: Creating the Safety Observation Process 68

9.1 How Do You Create the Observation Process? 70
9.2 Analyze Past Incidents and Injuries 70

CONTENTS vii

9.3 Develop a List of Critical Safe Practices 71
9.4 Draft and Revise Checklists 77
9.5 Develop the Observation Procedure 82
9.6 Feedback on Observations 88
9.7 Trial Run the Observation Checklist and Process 92
9.8 Conduct Management Review 93

CHAPTER 10 Step 3: Designing Feedback and Involvement Procedures   94

10.1 Develop Guidelines for Using Graphs 95
10.2 Plan Reviews of Safety Process Data 98
10.3 Develop Guidelines for Setting Improvement Goals 99
10.4 Establish Guidelines to Expand Involvement in Observations 100
10.5 Checklist for Planning Feedback and Involvement 101

CHAPTER 11 Step 4: Developing Recognition and Celebration Plans   102

11.1 Overview of Safety Awards and Incentives 104
11.2 Safety Recognition 104
11.3 Simple and Concurrent Safety Awards 106
11.4 Tiered Safety Awards 108
11.5 Support through Traditional Compensation 117
11.6 Safety Incentive Compensation 117
11.7 General Guidelines on Supporting Safety Motivation 119

CHAPTER 12 Step 5: Planning Training and Kickoff Meetings   120

12.1 Observer Training 121
12.2 Plan Kickoff Meeting(s) 122
12.3 Plan Training Needed to Support the Process 123

CHAPTER 13 Step 6: Conducting Management Review   125

CHAPTER 14 Implementing Behavioral Safety Process   128

14.1 Conduct Training for Steering Committees 129
14.2 Establish a Process Owner 130
14.3 Steering Committee's Responsibilities 130
14.4 Management's Responsibilities 134

CHAPTER 15 Maintaining the Behavioral Safety Process   135

15.1 Steering Committee Members' Responsibilities 136
15.2 Common Situations 138
15.3 Steering Committee's Responsibilities 138
15.4 Management's Responsibilities 141

CHAPTER 16 Some Final Suggestions on Implementation   143

CHAPTER 17 Special Topics: Safety Leadership   145

17.1 Biggest Barrier to Effective Safety Leadership   145
17.2 Other Barriers to Effective Safety Leadership   146
17.3 Leadership's Special Role   147
17.4 Phases of Management Support   148
17.5 Management's Most Important Role   149
17.6 Positive Questions   150
17.7 Formal Monitoring and Management Action Items   151
17.8 Role of Managers and Supervisors in Observations   152
17.9 Importance of Informal Leaders   153
17.10 Other Leadership Responsibilities   154
17.11 Safety Leadership Checklist   154
17.12 Concluding Comments on Leadership's Role   155

CHAPTER 18 Special Topics: Serious-Incident Prevention   157

18.1 Element 1: Build Management Commitment and Leadership   159
18.2 Element 2: Involve Employees   159
18.3 Element 3: Understand the Risks   160
18.4 Element 4: Identify Critical Work for Controlling the Risks   160
18.5 Element 5: Establish Performance Standards   160
18.6 Element 6: Maintain Measurement and Feedback Systems   161
18.7 Element 7: Reinforce and Implement Corrective Actions   161
18.8 Element 8: Improve and Update the Process   162
18.9 Implementation of the Serious-Incident Prevention Process: Pipeline Operations Case Study   162

CHAPTER 19 Special Topics: Self-Observation Process   169

19.1 What's My Job?   170
19.2 How Am I Doing?   170
19.3 What's In It for Me?   171
19.4 How to Implement a Self-Observation Process?   171
   Select a Safety Representative from Each Work Group   171
   Create an Index   172
   Develop a Sampling Process   173
   Post the Self-Observation Data   174
   Provide Group and Individual Recognition   175
19.5 Final Suggestions on Self-Observations   177

CHAPTER 20 Special Topics: The Steering Committee   178

20.1 Creating the Steering Committee   178
20.2 Training the Steering Committee   179

CONTENTS ix

20.3 Steering Committee Responsibilities 179
       A. Managing Process Measures 179
       B. Managing Behavioral Safety Measures 183
       C. Managing Safety Results or Outcome Measures 183
20.4 Responsibility Summary 186

CHAPTER 21 Advanced Topics: Why It Works and Behavioral Basics 188

21.1 Pinpointing 188
21.2 ABC Analysis 190
21.3 Consequences 191
21.4 Antecedents 194
21.5 Individual Learning History 195
21.6 Behavioral Analysis Worksheet 196
21.7 Developing an Action Plan to Address Behavioral Causes 198

CHAPTER 22 Advanced Topics: Improvement Projects 199

22.1 Problem-Solving Steps 199
22.2 Methods of Gathering Additional Information 200
22.3 Identifying Weak or Missing Contingency Elements 202
22.4 Guidelines for Setting Goals 204
22.5 Guidelines for Recognition and Celebrations 205

CHAPTER 23 Other Support Programs 208

23.1 Additional Safety Process Components 208
23.2 Supplemental Safety Programs 208
23.3 Additional Safety Process Components 209
23.4 Common Support Programs 211
23.5 Coordinate Special Programs 214

CHAPTER 24 Long-Term Case Studies 216

24.1 Behavioral Safety in a Refinery 216
       Phase 1: Pilot Area 216
       Phase 2: Plantwide Implementation 218
24.2 Employee Safety Process at an Ore-Processing Facility 219

CHAPTER 25 Self-Observation Case Studies 223

25.1 Canadian Gas Production and Pipeline Company 223
25.2 Electric Utility 227
25.3 Logging Industry 229

CHAPTER 26 Small-Company Case Studies 234

26.1 Pipeline Company 234

26.2   Polyolefin Plant   235
26.3   Food-Processing Plant   237

CHAPTER 27   Observer Effect   **240**

CHAPTER 28   Original Case Studies   **253**

28.1   Employee Safety Process at a Gas Pipeline Company   254
28.2   Creating a Positive Safety Process   255
28.3   Lessons Learned in Enhancing Safety Performance
       in a Paper Mill   256
28.4   Behavioral Approach to Industrial Hygiene   259
28.5   Long-Term Effects of a Safety Reward Program
       in Open-Pit Mining   266

Appendix A: Sample Implementation Schedules   **271**

**Appendix B: Selected Consultants Experienced in Implementing
             Behavioral Safety Processes**   **274**

**Appendix C: Unstructured Approach to Identifying and
             Defining Values**   **275**

Clarifying your Values   275
Phase 1: Identify Your Basic Values   276
Phase 2: Pinpoint Practices That Exemplify Those Values   276
Phase 3: Provide Training on the New Values   277
Phase 4: Use the Values as Basic Ground Rules for Interactions   278

References   **279**

Index   **283**

FOREWORD

One of my primary professional goals has been to encourage people to actively care for the safety and well-being of others in their communities and work settings. For the past two decades I have been traveling around the country promoting the concept of actively caring in both communities and organizations. In the context of organizations, actively caring is defined as employees acting to optimize the safety of other employees (e.g., giving rewarding feedback to an employee working safely or giving corrective feedback to a co-worker working at risk). That's what *Values-Based Safety Process: Improving Your Safety Culture with Behavior-Based Safety* is all about. It teaches both the mechanics of how to design and implement a behavioral safety process in any organization and the importance of creating an environment where everyone actively cares about his or her fellow employees or associates.

In discussing the importance of values, Terry McSween embraces my concept of actively caring while also addressing a number of other cultural practices that are crucial to the success of a behavioral safety process. Many of these elements are quite familiar to those involved in total quality efforts. For example, he emphasizes the importance of eliminating blame from the workplace. He suggests that data be used for problem solving, not victim blaming. He also cautions us about using data from the safety observation process (i.e., "percent safe") as a basis for personnel evaluations for either individuals or supervisors, suggesting that such practices will destroy the integrity of the process. In addition, he warns us of the dangers of mandating standards for such observational data, again because such practices will destroy the integrity of our improvement efforts.

Dr. McSween shows the reader how to plan and implement a behavioral safety process. All of us realize the difficulty of developing a cookbook implementation, but he clearly describes what must be done while including potential design options and issues to consider when selecting from those options. In short, while I do not believe anyone will ever write the perfect safety cookbook for managing safety, this one is as close as I have seen.

In the safety field, it is rare to find well-written books with straightforward, practical, and effective guidelines for developing companywide action plans. This is one of those books. Read it carefully. It will help you make a beneficial difference in your organization.

E. Scott Geller

Virginia Polytechnic Institute and State University

PREFACE

When the first edition of *Values-Based Safety Process* was written in 1993, my goal was to illustrate a proven approach to applying behavioral techniques to improve safety in the workplace. Since that time the field and interest in behavioral safety have greatly expanded, as has our experience. With that in mind, I set out with three primary objectives for the second edition: bring the methodology and terminology into alignment with current practice, address a variety of special topics, and provide case studies that more adequately represent current practices and application.

In this edition, I resequenced the entire implementation process and use the *design team* and *steering committee* terminology more consistent with current practice. In Chapter 9, I have greatly expanded the section on creating observation checklists, adding both additional worksheets and added current examples.

The reader will also find new chapters addressing special topics. In particular, the role of leadership (Chapter 17), addressing serious incidents (Chapter 18), employees working in isolation (Chapter 19), and the role of the steering committee (Chapters 20 and 22) all receive in-depth coverage.

This edition also includes eight new case studies. The new case studies show the long-term effectiveness of a behavioral approach (Chapter 24), illustrate the effectiveness of a self-observation process (Chapter 25), and document the effectiveness of behavioral safety in smaller organizations (Chapter 26). Finally, an invited chapter from Alicia Alevero and John Austin presents their research that demonstrates the positive impact that conducting observations has on the *observer*! This research on what they have entitled the "observer effect" is important because it documents the value of getting employees involved in conducting safety observations. I am excited to be able to include a summary of their research in Chapter 27.

TERRY E. MCSWEEN

ACKNOWLEDGMENTS

I owe a debt of gratitude to my teachers and mentors and to the behavioral researchers who originally developed the ideas presented in this book. In particular, Richard W. Malott of Western Michigan University has been, and continues to be, my teacher, manager, coach, and friend. Several friends and customers also deserve special thanks. Among them, Rixio Medina of CITGO Corporation in Tulsa, Oklahoma, has been a special supporter and friend for many years, and Kem McVey, who provided my first opportunity.

For their efforts in preparing this second edition, I thank my associates at Quality Safety Edge: Tom Burns, Grainne matthews, Wanda Myers, Ann Pinney, and Judith Stowe. They provided critical assistance with the new chapters and case studies included in this edition.

I reamin indebted to the people who assisted in editing the first edition of this book: Dr. Maria Malott, Malott & Associates, Kalamazoo, Michigan; Dr. Dale Brethower, Western Michigan University, Grand Rapids, Michigan; and Dr. Victor Zaloom, Lamar University, Beaumont, Texas. In addition, I greatly appreciate the work by Beth Sulzer-Azaroff, Bill Hopkins, Tracy Thurkow, and Gim Getting in preparing three of the case studies included in the first edition.

Also, I am particularly indebted to J. K. Hillstrom of Houston, Texas, for his technical writing and editing skills that greatly simplified my task and kept me focused on the reader in both editions.

Finally, this edition would not have been possible without the support and sacrifice of my wife, Camie, and son, Daniel. They bring new meaning to my life.

1 Safety Basics

Improving safety is often difficult, partly because of past success. Most businesses and industries today have excellent safety records. Based on the year 2000 average of three lost-workday cases per 200,000 work-hours, the average employee in American industry can expect to experience one lost-workday injury in approximately 33 years of work. The problem with this high level of safety is the complacency that it often creates. Employees can shortcut a safety procedure, yet rarely get hurt. Behaviorally, therefore, many of them view the slight probability of suffering an injury as not significant enough to maintain 100 percent compliance with safety procedures. The reduced likelihood of an injury often simply does not offset the immediate comfort, convenience, or time saving associated with an unsafe shortcut. But the behavioral results are predictable for many companies. The overall frequency of unsafe acts remains too high and safety incidents that include serious injuries continue at a statistically predictable rate.

This chapter reviews what most organizations do in dealing with behavioral safety issues.

1.1 TRADITIONAL SAFETY PROGRAMS

Most companies have embraced the following programs and initiatives to improve compliance with safety procedures:

- Informal feedback on complying with safety procedures
- Safety meetings and training
- Safety awards
- Safety audits
- Written procedures
- Special initiatives (posters, newsletters, off-the-job safety programs, etc.)

All these procedures are important to a successful safety process. Done properly, these elements contribute to good safety performance. But today, these elements define average safety efforts—they are what everyone does. If a company does them well, it will achieve an average level of safety for the industry. Although the rate of injuries will be affected by the consistency of these efforts, the result will basically be normal variation above and below the industry average: some years better than average, some years worse.

2 SAFETY BASICS

Consistent safety excellence requires far greater consistency in how safety is managed than most companies achieve through traditional methods. Research by Du Pont and others suggests that 80 to 90 percent of today's incidents are a result of unsafe acts rather than unsafe conditions. Thus, very few companies that focus on the latter achieve consistently high levels of compliance with their safety procedures.

New research confirms the effectiveness of a behavioral approach to safety that increases compliance and greatly reduces incidents. This book presents the kind of results that can be achieved through a behavioral safety approach, a summary of the key components of a behavioral safety process, and an overview of the procedures for implementing a behavioral approach within an organization's existing safety efforts.

In each of the cases described below, the companies had previously used the elements of traditional safety programs. In one example of improvement, on changing in 1980 to a behavioral approach, a major U.S. drilling company reduced its Occupational Safety and Health Administration (OSHA) recordable injury rate by 48 percent and moved from the industry average to being one of the industry's top five safety performers. This improvement was achieved through a management-driven behavioral approach even without the levels of employee involvement typical of current implementation efforts (Fig. 1.1).

In another case, on adopting a behavioral approach, a solids-handling chemical company with incident rates more or less typical of most such companies at the time went from three or four OSHA-recordable injuries per year to no recordable injuries over a period of more than 18 months (Fig. 1.2). This was a union plant and the hourly employees initiated the new approach, stating they were "tired of being beat up because of safety." They wanted to create a positive safety process that was employee driven.

Finally, a division of a large pipeline company achieved zero injuries for three years, a vast improvement over the prior six years (Fig. 1.3). This company initially planned to implement a self-observation process but during the planning found it

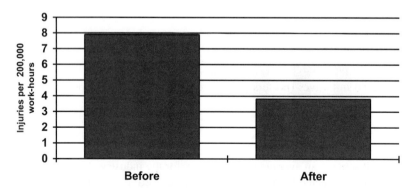

Figure 1.1. Reduction in recordable injuries achieved by a drilling company using a behavioral safety process.

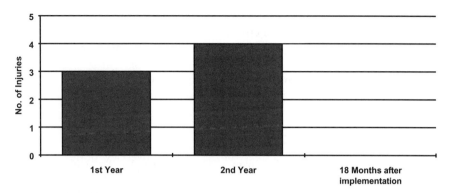

Figure 1.2. Reduction in recordable injuries achieved by a chemical plant using a behavioral safety process.

could schedule employees in a way that allowed peer observations in the field. The latter proved to be highly effective.

The process that achieves these results is well documented by both experimental studies and direct experience. The key components are basic:

- A behavioral observation and feedback process
- Formal review of observation data
- Improvement goals
- Recognition for improvement and goal attainment

These elements appear so simple and common sense that many people underestimate the difficulty involved in creating a behavioral safety system. Managers, in particular, often fail to anticipate the difficulty in achieving the level of consistency and support required to make the approach successful. However, these elements

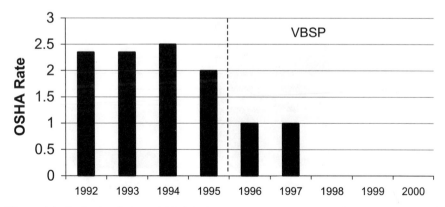

Figure 1.3. Reduction in recordable injuries achieved by a division of a pipeline company using a behavioral safety process.

4 SAFETY BASICS

combine to provide a proven process for systematically managing safety on the job in a way that minimizes the risk of error due to unsafe acts, ensures a high degree of procedural compliance, and maintains that level of performance consistently over extended periods.

Before examining the behavioral safety process in more detail, let us take a closer look at some of the key elements of current safety improvement efforts.

1.2 DU PONT'S SUCCESS

In colonial days, the Du Pont Company made black powder. The Du Pont family planned and built their factory into a hillside in a way that would direct the force of an explosion out over the Delaware River. This orientation protected the workers' homes and families in the village located behind the factory. The risks in this business meant that they had to think about safety all the time: Their lives depended on it.

Today, Du Pont continues to place a heavy emphasis on safety. The company continues to promote innovation in industrial safety. Over the years it has been among the first to champion the following, among other safety management practices:

- Layered safety audits
- Safety audits focused on behavioral instead of environmental factors
- Specific feedback techniques during audits

These additional elements of Du Pont's approach to safety evolved from a formal study of all lost-workday cases that the company experienced over a 10-year period. The results of this study suggested that 96 percent of Du Pont's injuries resulted from unsafe acts rather than unsafe conditions (Fig. 1.4). Their study supported

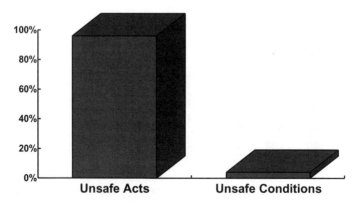

Figure 1.4. Du Pont found that 96 percent of its lost-time accidents over a 10-year period resulted from an unsafe act by an employee.

findings from 1929 that suggested 88 percent of all injuries were a result of unsafe actions by employees rather than unsafe conditions (Heinrich, 1959). Du Pont's data lend credibility to Heinrich's work, even though various authors later criticized his methodology.

Based on these results, Du Pont refined its approach to safety into its present Safety Training Observation Program (STOP). Du Pont promotes STOP extensively both within and outside the company. STOP involves a process of *layered safety audits* in which each layer of management conducts a regular safety audit, typically every week. A manager enters an area and finds its superintendent; then they conduct a safety audit of that area. On a different week, the superintendent chooses an area supervisor and they conduct a safety audit. Further, all management personnel conduct a formal audit each week in one of the work areas for which they are responsible while also conducting informal observations of both safety practices and safe work conditions at all times.

As they conduct the periodic audits, managers and supervisors complete STOP cards to document any unsafe acts they have observed, though not documenting the names of the observed employees. However, as soon as convenient, they approach an employee who performed an unsafe act and ask two questions. The first is a "What could happen?" question that prompts the employee to identify which of the observed actions created the risk of an incident. The second is a "How could [the employee] do the job safely?" question that prompts the employee to identify how to do just that.

Along with STOP, Du Pont strongly emphasizes the importance of safety in many other ways. The company has extensive safety training materials to support safety meetings and planning, such as Take Two, a safety program that encourages employees to take 2 minutes to consider the safety aspects of each job before beginning work. It tracks off-the-job injuries and conducts formal off-the-job safety programs and training. Also a formal procedure, any lost workday due to a safety incident prompts a site visit from an executive of the company who personally reviews the incident investigation and interviews all personnel involved in the incident. Informally, employees' safety records follow them throughout their Du Pont careers. These elements combine to create a "safety culture" that routinely results in the safety performance shown in Figure 1.5. Du Pont is usually number 1 in safety in the chemical industry, and historically it has frequently been twice as good as the next safest company.

Although Du Pont's safety record is very good, the average for the entire chemical industry is also very good. The industry average represented in Figure 1.5 means that a chemical plant employee has a very low probability of getting hurt. On the basis of chance, a chemical industry employee will suffer an injury incident on the average of once in every 30 years of work. By the same measure, a Du Pont employee will suffer an injury requiring medical treatment on the average of once in every 100 years.

These low probabilities of injury comprise part of what makes further safety improvements such a challenge. We will provide additional discussion of the behavioral impact of these probabilities later in the book.

6 SAFETY BASICS

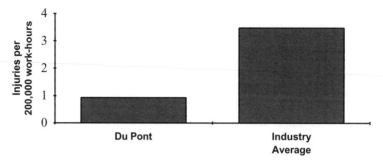

Figure 1.5. Du Pont has an exemplary safety record in the chemical industry (data for 1991).

1.3 OUR FINDINGS

Our studies replicate the Du Pont findings regarding the extent to which unsafe behavior contributes to injuries. Over the past 10 years, we have analyzed injuries at hundreds of organizations in developing checklists to help prevent injuries. Our findings suggest that in most organizations behavior contributes to between 86 and 96 percent of all injuries. Figure 1.6 presents data from one of these studies

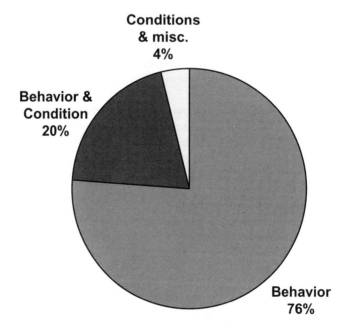

Figure 1.6. This data from a representative company shows the extent to which behavior contributes to injuries.

that replicate Du Pont's findings of behavior contributing to 96 percent of all injuries.

These data are not meant to suggest that employees are directly to blame for 96 percent of their injuries. From the perspective of behavioral psychology, all behavior is a function of the environment in which it occurs. Unsafe work behavior is accordingly the result of (1) the physical environment, (2) the social environment, and (3) workers' experience within these. The remainder of this book is dedicated to how to change the workplace environment in ways that increase safe behavior.

Several other lessons were also learned from these analyses. First, when we examined serious injuries and fatalities, we found them almost always in the category of "behavior and conditions." That is, serious injuries and fatalities most often result from a combination of unsafe behavior and unsafe conditions. Safety professionals often talk about a chain of events leading to an injury. Some of the links in the chain are behaviors, some are conditions, and we can often prevent injuries by breaking any of the links.

One story clearly exemplifies this combination of factors. A 40-year-old coker unit at a major refinery experienced a train derailment early in the day that resulted in a grate being removed from a walkway across railroad tracks, thereby exposing a pit about 5 feet deep. (Coker units are huge facilities from which finished coke is often unloaded directly into rail cars.) Workers immediately placed barricades on the walkway on each side of the railroad tracks. Some hours later, the coker unit had a pump failure that caused the area to be flooded with boiling hot water. When the shift changed later in the day, several of the incoming employees were not adequately briefed on the condition of the unit and did not learn about the derailment and the resulting pit now hidden by the still very hot water. Three of these employees began to walk through the area and came upon the barricades. Seeing the water, two of the three walked around the area. The other employee was wearing knee-high rubber boots. Assuming the barricades were intended simply to prevent employees from walking through the water, he walked around the barricade, stepped between the railroad tracks, and fell to his chest in boiling hot water. He was off the job for over 14 months as a result of his burns.

Clearly, a combination of factors contributed to this event, beginning with the unsafe conditions created by the train derailment and pump failure. A number of behaviors also contributed to it. The barricading was inadequate after the area flooded. The incoming employees were not adequately briefed about the condition of their unit. The employee who was injured walked around one of the barricades, and his co-workers allowed him to do so. In addition, management had cut an item from the capital budget the previous year for an upgrade that would have prevented the area from flooding. Instead of funding the upgrade, management had arranged a Band-Aid solution by building a platform that gave operators access to valves in the area when flooding occurred. Accordingly, one could wag a lot of fingers, but the point is that breaking any of the links in this chain of events could have prevented this very serious injury from occurring.

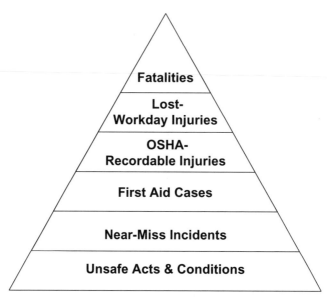

Figure 1.7. Safety triangle shows relationship between unsafe work situations and injuries.

1.4 THE SAFETY TRIANGLE

Du Pont's emphasis on unsafe acts recognizes the hierarchy commonly represented by the safety triangle in Figure 1.7. Geller (1988) refers to this as the reactive triangle. He suggests that approaches based on this model are typically reactive rather than preventive in that the focus is on decreasing unsafe acts. He maintains that a better preventive approach must focus on increasing and maintaining safe acts.

Additional support for Geller's logic comes from a study by Reber and Wallin (1984). Using an observational procedure in a heavy manufacturing environment, they reported a significant negative correlation between percentage of safe behaviors and both the rate of injuries and the rate of lost-time injuries. In other words, the results of their study showed that the lower the rate of safe behavior, the higher the rate of injuries. Their data suggest that increases in safe behavior should result in lower incident and injury rates. The empirical studies discussed in the sections that follow have confirmed the effectiveness of such a preventive approach.

1.5 COMPLACENCY

Complacency refers to the loss of the fear of injury that typically motivates employees to work safely. When we talk of wanting people to perform their jobs safely for the right reasons, we usually mean they should work safely to avoid the pain, suffering, and lost wages associated with injury. Too often the problem is that employees become complacent and begin to shortcut safety procedures. The safety triangle in Figure 1.7 also helps explain this complacency, a prevailing problem when trying to promote safety.

If the probability of getting injured is high, complacency is not a problem. For example, we seldom have a problem in getting welders to use appropriate eye protection. The probability of burning their eyes is great enough that they are very consistent about wearing eye protection.

The implication of the safety triangle is that many unsafe acts and conditions occur before an injury results. The frequency of these events reflects the probability at each level of the triangle. The probability of getting injured is often simply too small to sustain a consistent level of safe work practices. Each time employees shortcut a safety procedure and do not get hurt, they lose a bit of the fear that typically motivates safety.

Clinical psychologists use a technique called *systematic desensitization* to help people overcome phobias, or irrational fears of such things as flying, snakes, spiders, heights, and so forth. The process involves gradual exposure to the feared item or activity. The same process seems to occur in the natural environment, and it works against us with regard to safe work habits. When an employee first works at heights, for example, the fear of falling provides strong motivation to consistently use appropriate fall protection. After several years on the job, the employee often has a much greater level of comfort and much less fear of falling and is therefore more likely to work without appropriate fall protection.

The behavioral safety process is designed to offset this phenomenon.

1.6 SAFETY AS A TEAM PROCESS

In an effort to promote safety and fight such complacency, many organizations establish initiatives to get employees involved in safety improvement, often using the same approaches taken in their quality improvement efforts. A natural outgrowth of these efforts is the involvement of employees in teams that work on improving safety. Such teams are directed to identify safety problems as well as develop solutions for them.

These teams give employees greater control over the types of improvement efforts initiated in their work areas. The advantage of a team approach is the increased ownership and support that team members have for programs they design and initiate. Shifting some of the accountability for safety initiatives from management to employees also places such responsibilities closer to the job. Further, it reduces the implied threat of punishment often found in mandated safety programs, resulting in a more positive work environment.

However, team approaches to safety typically struggle with several common problems. One is that teams often shift their priorities as they begin to feel they have safety under control. The result is a cycle of incidents. This problem is frequently a characteristic of team safety improvement efforts whether the safety teams are comprised of managers or employees. After successfully reducing the

10   SAFETY BASICS

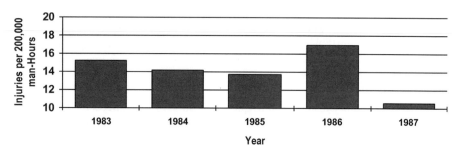

Figure 1.8. A cycle of accidents often characterizes safety improvement efforts.

rate of incidents, attention goes elsewhere and the likelihood of injuries increases. Then the teams must readdress safety.

Figure 1.8 presents data from a plant with safety teams involved in a continuous improvement effort. Managers and supervisors also participated, and these teams were given responsibility for continuous safety improvement in different work areas of the facility. The graph shows the success of these teams and the characteristic cycle of injuries that invariably prompts renewed safety improvement efforts. In some cases, such cycles are simply normal variations that characterize companies using traditional safety programs. In other cases, the variances correlate with inconsistent levels of attention being paid to safety.

Employee teams also experience other problems related to the design of an organization's team process. Often, teams spend inordinate amounts of time identifying safety problems and, in doing so, frequently generate either suboptimal solutions or new problems for others to resolve. The additional problems are usually the result of a poorly designed team process that fails to achieve the following:

- Ongoing communication with management
- A structured process for selecting problems and developing solutions
- A good understanding of how to deal with human performance issues

Employees are seldom familiar with proven methodologies such as those suggested in behavioral research studies. Without specific training, they do not have the knowledge needed to develop the behavioral systems and procedures that will improve safety compliance. They generally merely implement more of the kinds of programs they have seen in the past. The team may look into safety posters or spend considerable time on safety awards. Or it may focus only on conditions and propose expensive physical changes of questionable cost-effectiveness or else simply compile multipage lists of safety items needing attention.

Depending on the industry, our experience suggests a team approach to safety can achieve an incident rate in the range of 5 to 10 recordable incidents per 200,000 work-hours. With frequent management attention to safety and a high degree of consistency in using traditional approaches to safety, many organizations do much better than that. However, the achievement of consistently high levels of safety, year after year, requires a behavioral observation process such as the one described in chapters that follow. Implementing behavioral safety through a team approach is the best way to ensure a consistently high level of attention to safety. It is also the most effective way to involve employees in developing and maintaining safe work practices.

1.7 COMMON PROBLEMS WITH SAFETY EFFORTS

Many contemporary safety improvement efforts suffer a further common set of problems:

- Employees may suffer severe consequences for reporting incidents.
- Safety awards are not related to behavior on the job.
- Management or staff make all plans and decisions regarding safety.
- Organization relies on punishment to reduce unsafe acts.

Employees may suffer severe consequences for reporting incidents. One problem in designing an effective safety program relates to the maxim, "Don't shoot the messenger." Severe penalties for experiencing incidents and significant incentives for not having incidents encourage employees not to honestly report minor incidents and injuries. This is a basic problem with formal or informal policies that penalize managers or employees with potential loss of income or promotional opportunities if they report an injury. In such an environment, even the well-intentioned supervisor may listen to an employee's claim that "It's really nothing serious. I wouldn't think twice about an injury like this at home." Or, especially if the injury is not likely to require time away from work, an employee may simply report that an injury occurred at home rather than at work.

Employees can feel similar pressure not to report minor incidents if safety awards are significant, particularly if the system provides individual awards based on their group's performance. Such incentive programs contribute to peer pressure that can encourage false reporting, thereby giving the appearance that the rate of recordable injuries is going down. Similarly, a steady rate of lost workdays is a red flag indicating that recordable incidents actually remain unchanged despite statistical improvements reported by the organization.

Whether or not an employee gets a Band-Aid and creates a first-aid case is not important. What is important is that the organization has a chance to learn from an incident that could have had more serious consequences. If a situation that caused an employee to need a Band-Aid could have resulted in a crushed hand, the organization must have a process for documenting and learning from the minor incident. Poorly designed awards programs and the threat of disciplinary action can jeopardize the integrity of systems designed to document such events.

Unfortunately, some less than knowledgeable consultants continue to recommend the use of such safety incentives in ways that may expose their clients to added liability. One of our prospective clients had set up an incentive program that provided a $25 bonus for employees of work groups that went four months without an injury and 4 hours off with pay for employees in work groups that went six months without a recordable injury. Predictably, the rate of recordable injuries went down while the rate of lost-time injuries remained unchanged. In short, the incentive program reduced the reporting of minor injuries but did not encourage employees to work more safely.

The risk of such simplistic approaches is increasing because of recent court decisions. A few years ago, an employee in Texas won a workmen's compensation case against an employer who had a safety "bingo program" that discouraged employees from accurately reporting injuries.[1] In another case, a court ruled that a cash incentive that discouraged employees from reporting on-the-job injuries could provide evidence that an employee was terminated for filing a workmen's compensation claim on an incident that he had not reported in accordance with company policy.[2] Depending on other aspects of a company's loss prevention efforts, an incentive that even inadvertently encourages employees to hide incidents can increase a company's liability when an employee has an injury. Such a safety incentive system certainly increases an employee's ability to file a compensation claim successfully well after the typical time period allowed for such claims.

Safety awards are not related to behavior on the job. A related problem with safety awards is that most do not reinforce safe behavior on the job. A typical program may base awards on individual or team performance. If individual performance is the basis for safety awards, an employee usually earns an award by working for a month, a quarter, or a year without an injury. If group performance is the basis for safety awards, the group's employees earn awards by working similar periods without an injury to anyone in the group. Companies must track injuries for regulatory purposes, and it is easy to base awards on these data. The result is that many organizations base their awards on outcome measures such as safety statistics instead of on process measurements and behaviors that promote safety.

The problem with this approach is that too many employees simply roll the dice; they take shortcuts that allow them to complete jobs more quickly and comfortably. With today's low incident rates, they will probably not have an injury. Meanwhile, they work with other employees who follow safety procedures day in and day out: employees who always wear a safety harness above 4 feet, employees who always get appropriate permits, and employees who always wear appropriate protective equipment. At the end of the year, however, these employees all get the same safety award. In fact, the likelihood of getting the award is roughly the same as the likelihood of getting hurt, so an employee in a typical U.S. company that focuses on such awards will get one an average of 32 out of every 33 years. Such awards are

---

[1] *Paragon Hotel Corporation v. Ramirez*, El Paso Court of Appeals, 783 Southwest 2nd 654, 1990.
[2] *Glass v. Amber Inc.*, Texas Court of Appeals no. 01-00-00589, 2001.

little more than gifts to both safe employees and those who take risks. Such gifts may promote safety awareness, but they do very little to motivate employees to work safely on the job.

Management or staff make all plans and decisions regarding safety. A further problem with many safety programs is that they are management based. While they may assign hourly employees to participate on committees responsible for safety programs, most companies rely on managers and supervisors to enforce safety rules and procedures. Many have some version of the layered safety audits championed by Du Pont. Management has the responsibility for discipline and must, therefore, be responsible for taking corrective action with employees—which the latter invariably regard as punishment. The result is that employees rely on managers to ensure safety instead of watching out for one another. Employees are not challenged to achieve safety improvements and do not get a sense of accomplishment from such improvements in their work areas. In short, for many employees in these organizations, someone else has the primary responsibility for safety. The process does not effectively create active, ongoing involvement for most employees.

Organization relies on punishment to reduce unsafe acts. A related problem is that many safety programs are primarily punishment based. As stated above, punishment-based programs are reactive in focusing on reducing unsafe acts rather than preventive in focusing on encouraging safe acts. An unsafe act cannot be corrected until it occurs, however, and each unsafe acts place employees at an increased risk of an injury. Safe acts, on the other hand, reduce the risks of an injury. Furthermore, a reliance on punishment creates additional problems that often are not readily apparent.

1.8 PROBLEMS WITH PUNISHMENT

Many of the problems just discussed relate to a traditional reliance on using punishment and a "corrective action" approach to safety. Most of us have used punishment for a very simple reason—it works. But because it works quickly, its immediate payoff often causes some individuals, particularly those in positions of power, to become accustomed to its use. This is seen in relationships at home, in our schools, and at work. At home, when children are making too much noise, a parent may respond by yelling, threats, or even physical punishment. The immediate result is that the children are quiet, thus resolving the matter for the parent. This immediate result also reinforces the parent's use of punishment. Because such positive outcomes immediately and consistently follow the use of punishment, the parent may begin to use it more frequently. The same dynamic occurs in other settings. The husband who criticizes his wife and the supervisor who criticizes an employee both get an immediate result that reinforces their use of punishment and criticism. Too often, these may begin to characterize the relationship.

On the other hand, when we use positive feedback, the outcome is often less immediate and less clear. What happens when a supervisor provides positive feedback for following a safety procedure or when a parent provides positive feedback to a child who is doing homework? The resulting behaviors may be more positive on future occasions, but the outcome does not have the same immediate impact as punishment or corrective feedback. This combination of factors often causes relationships to get out of balance with too much negative feedback and too little positive feedback.

The overuse of punishment has several further disadvantages, whether in the home, the schools, or the workplace. Among these, punishment

- must be either severe or highly probable,
- is effective only in the presence of the punisher,
- often teaches the wrong lesson,
- damages relationships and involvement,
- runs contrary to the philosophy of our quality efforts, and
- is difficult to maintain.

It must be either severe or highly probable. Research suggests that punishment must be either severe or highly probable to be effective (Malott, Malott and Trojan, 2000). People behave as they do for a reason. If they are behaving in a given way, something is maintaining that behavior. In the workplace, people will ignore safety requirements for comfort or convenience. If punishment for doing so is not consistent and severe enough to be significant, it will not offset the natural incentives they have to ignore the safety requirements. The use of significant punishment also has other serious problems. Many companies have adopted policies stipulating that working safely is a condition of employment. The implication is that employees who take safety risks are risking their jobs. Too often, such policies have two undesirable effects: (1) they simply discourage employees from reporting minor incidents and near misses and (2) they discourage frank discussions of factors that contributed to an injury.

Similarly, if the risk of being caught is low or if the risk of significant corrective action is low, punishment will not offset the natural incentives that maintain unsafe practices in the first place. Employees will simply take their chances. In addition, in today's economy, most companies cannot afford to maintain a large management staff to ensure compliance with safety procedures. A police force approach to safety enforcement is no longer a cost-effective management strategy.

It is effective only in the presence of the punisher. An additional problem with reliance on punishment is that employees soon learn they must follow procedures only when certain supervisors or managers are present or nearby. They learn when they can bend the rules without fear of correction or discipline. The adage "When the cat's away, the mice will play" directly pertains here. Our educational systems often struggle with this problem. When a teacher who relies on punishment as the primary motivation for good behavior leaves the classroom for a length of time, more often than not the quiet and orderly classroom quickly becomes chaotic.

Many companies today have reduced the numbers and levels of management and hence can no longer provide the level of supervision they once maintained. The result is that they can no longer rely on a police force philosophy of safety. Although managers play an essential role in maintaining company policy and procedures, they must increasingly strive to encourage self-management and a shared responsibility for all aspects of safety.

It often teaches the wrong lesson. Punishment often teaches how to avoid being punished instead of the desired behavior. Thus we learn to be quiet when the teacher is present. We learn to generate reasonable excuses to explain why we act in a certain way. Overreliance on punishment can easily teach employees not to report minor injuries or near misses. In addition, it affects the accuracy of the information reported during incident investigations. An employee who gets something in his or her eye may claim to have been wearing safety glasses. If employees fear punishment, they will not report information accurately.

It damages relationships and suppresses involvement. An additional problem with punishment is that it damages relationships. People dislike those who routinely criticize or punish them. Punishment also often generates a great deal of emotional behavior that leads to counterproductive reactions, such as counterattacking or simply avoiding the punisher. In addition, people often find ways to get even with those who punish them, as suggested in the adage "What goes around comes around." Someone who is criticized or punished in a meeting will usually find a way to retaliate even before the meeting ends. In the workplace, punishment often leads to slow work or poor workmanship. Or it may lead to an employee doing exactly what the supervisor told the employee to do even when not appropriate, a reaction commonly known as "letting the boss hang himself (or herself)." Punishment destroys relationships that are the bedrock of teamwork.

Another major problem with overreliance on punishment is that it discourages cooperation and problem solving. When someone criticizes our work, do we go into a problem-solving mode? Do we work diligently to help resolve the problem? *No way!* What do we do? We become defensive. We begin to make excuses, to rationalize, to explain why we could not have done differently, perhaps to claim how those we depend on prevented us from behaving differently or otherwise blame others. Punishment undermines the cooperation required for teamwork.

Managers must instead create a work environment that encourages personal responsibility and minimizes blame. In environments that rely on punishment, damaged relationships are often a barrier to creating an effective partnership in which management and employees actively share responsibility for safety.

It runs contrary to the philosophy of quality efforts. These problems result in reliance on punishment going against the philosophy of current quality improvement efforts. Edwards Deming, one of the gurus of the quality movement, exhorted companies to "drive fear out of the workplace" and "remove barriers to pride in workmanship." Overuse of punishment is one of those barriers. In an environment motivated by fear, employees work because they have to, not because they want to. When we do something because of threats, nagging, or criticism, we seldom feel a sense of accomplishment or pride in the quality of our work performance.

It is difficult to maintain. Punishment-based efforts are difficult to maintain. Audits that may result in punishment are not fun for those doing the audits or for those being audited. Managers do not like punishing employees, and employees do not like being punished. Corrective feedback is almost always punishing, regardless of the method or severity of its delivery.

Organizations must learn not to rely on threats, nagging, criticism, or comparable forms of punishment to get work done safely. Management and employees need to form a partnership of shared responsibility for safety. Managers can no longer simply blame employees for safety problems. Employees cannot rely on managers to ensure their safety. Doing so is too dangerous.

1.9 APPROPRIATE USE OF PUNISHMENT

While an overreliance on punishment has serious problems, discipline does have an appropriate place in safety programs. Punishment works and works quickly. If some employees are doing something that endangers themselves or others, they must stop. If they are repeatedly violating a safety policy that they understand, they must change what they are doing to comply with the organization's safety requirements or realize that they are in fact making a career choice. Whether as corrective feedback or more stringent disciplinary action, punishment is often appropriate for such situations and acceptable to most employees.

Some authors, such as Geller (1997), suggest that disciplinary action is rarely appropriate in safety situations. Geller makes the case that employees should be disciplined only if they intentionally break a safety rule. The problem with this recommendation is that the only way we can know if a worker intentionally did something unsafe is to ask! Unfortunately, when faced with the prospect of significant disciplinary action for doing so, few workers are likely to admit they chose to commit an unsafe act.

In our society, we expect people to be responsible for both knowing and complying with applicable rules. For example, not knowing the speed limit is not a sufficient reason for avoiding a speeding ticket. We expect people to be conscious of their activities, whether driving at the speed limit or following safety rules. Accordingly, disciplinary action is appropriate for violating a lockout–tag-out requirement, for instance, even if the employee did not make a conscious decision to take the risk (assuming that the procedure is clear and that company has provided adequate training on its lockout–tag-out procedures).

As to safety, all employees must know their company's "rules of life" and understand that breaking any of these rules will result in disciplinary action. These are the rules that the company will always enforce with disciplinary action when necessary. They are also the rules that are critical to protecting people's lives. The line must be clearly drawn in the sand, and when employees step over the line, the result needs to be disciplinary action. Both employees and supervisors need to clearly understand these requirements. Further, supervisors and managers should understand that they too are subject to disciplinary action if they fail to consistently enforce basic safety rules.

To repeat, punishment is like a drug that produces an immediate high. It works, and it works immediately. Because punishment is so effective, we can easily get hooked on its use. In many of our relationships, both at work and at home, the immediate outcome can cause us to become overcritical and pay undue attention to problems and actions that irritate us. Our relationships can get pulled out of balance, and we may come to rely too heavily on criticism and corrective feedback.

Conversely, the rewards for using positive feedback are less immediate. We learn this difference at an early age. Imagine your child comes home from school with three A's and one C on the report card. What might be your first words? In safety, we have often seen the same thing—as in noting 10 hoses coiled and stored properly and 1 left strung across a walkway. How often do we acknowledge the appropriately stored hoses? We must strive for better balance, one that promotes better personal relationships and a more positive environment at work and at home. Doing so is especially important in achieving real and lasting safety in the workplace.

1.10 COMPONENTS OF A PROVEN SAFETY PROCESS

Fortunately, research conducted during the last decade can help us to identify the key components of a more positive system for addressing safety. Two noted researchers in the field of behavioral psychology, Judy Komaki and Beth Sulzer-Azaroff, have identified several primary features of an effective safety process. These two researchers and their associates have demonstrated and proved the effectiveness of the following components for improving safety:

- A behavioral observation and feedback process
- Formal review of observation data
- Improvement goals
- Reinforcement for improvement and goal attainment

In their studies, both researchers began by pinpointing safe behaviors in workplaces that would reduce the likelihood of incidents. They then set up an observation procedure that provided feedback and data regarding those behaviors. Next, managers reviewed the observation data, set improvement goals with the employees, and arranged reinforcement in the form of recognitions to celebrate successes in goal attainment. (The References at the end of the book provide a list of publications by these researchers and their colleagues.) [For an extensive review of the literature on behavioral safety, see Sulzer-Azaroff and Austin (2000).]

Researchers have not yet conducted extensive research into the importance of the individual components of these studies, though Komaki's (1986) research suggests that the observation process may be the most critical element of the package. Conducting research into the effectiveness of each of the components of behavior-based safety is difficult because each component contributes to only a part of the effectiveness of the intervention package. The studies that have investigated the individual components of this approach are included below.

Three studies by the original researchers have implications for the design of effective safety improvement efforts. One study proved that on-the-job feedback in conjunction with safety training produced a much higher level of compliance than training alone (Komaki et al., 1980). The second study examined the effects of worker participation in goal setting (Fellner and Sulzer-Azaroff, 1985). This study suggests that explicit goals improve the effectiveness of safety feedback and that whether supervisors or employees set those goals is not a critical factor. The third study examined whether supervisors or safety personnel should perform the observations (Fox and Sulzer-Azaroff, 1989). This study found no significant differences in safety performance related to whether supervisors or safety personnel conducted them and then provided feedback. The observation and feedback processes themselves were the critical factors.

Tom Krause is well known for his work in behavior-based safety. He documented the first long-term success of behavioral safety. His data showed that 73 companies were able to sustain and continue safety improvements for five years following implementation (Krause et al., 1999). This work is particularly significant because many of the early interventions appeared to be short lived, lasting only for the duration of the research study.

A recent series of studies suggests the importance of employees conducting safety observations. Austin and Alvero (in press, also Chapter 27) have shown that when employees conduct safety observations using a safety checklist, their performance of safety practices on the checklist improves and becomes more consistent. These studies lend credibility to our own data suggesting that (1) departments with high levels of observations have lower injury rates than departments that are not as successful at conducting observations and (2) the injury rates of employees who participate in conducting observations are 50 percent lower than those of employees who do not conduct observations.

In summary, studies have documented the long-term effectiveness of an intervention package that includes an observation process, behavioral feedback, improvement goals, and reinforcements for improvement. The empirical data on the components of this approach suggest that an observation process, on-the-job feedback, and improvement goals are each important to maximizing safety performance. Based on the data available when the first edition of this book was written in 1995, who conducted the observations did not appear to be a critical factor. The data now clearly suggest the importance of involving *all* employees in conducting safety observations.

2 The Vision

The following scenario illustrates a typical behavioral safety pocess. The description is generic and in actual application many of the details may vary depending on the needs of the organizaiton.

2.1 A TYPICAL WEEK

At the end of Monday's safety meeting, Randy, the maintenance group's employee safety representative, announces that on Tuesday he will conduct a safety observation of the maintenance crew working on Unit 1. On Tuesday afternoon Randy gets Jim, another maintenance employee, and says, "Come with me to Unit 1. I want to do a safety observation of the crew there and would like you to come see how these are done so that you can do them later on. Your participation is totally voluntary, but it'll help us ensure a safe workplace."

Jim agrees to go along. As they head for Unit 1, Randy and Jim discuss the fact that soon everyone in the plant, including the site manager and supervisors, will finish observation training and those who volunteer to do so will conduct safety observations. Everyone will be a partner in safety and share responsibility for achieving it.

When Randy and Jim arrive at Unit 1, Randy takes a detailed, one-page checklist out of a folder. Without referring to the checklist, the two men first scan the work area. After Randy explains the procedure to Jim, they ask themselves, "What do we see these employees doing that could cause someone to get hurt?" They note one such practice on the checklist. They further review the checklist, mark each safe practice, and check areas of concern. They do not record any names on the checklist.

After completing their observations, they approach the employees and review the checklist with them. Randy says, "We noticed you were all using the required personal protective equipment. Your work areas are neat, your tools are well organized, and you were using the right tools for the work you were performing. However, we also noted a lack of barricades to prevent personnel from passing through your work area." Randy and Jim answer a few questions, after which Randy asks the employees to rope off the work area. The two men then return to Randy's office where Randy puts the completed checklist into a three-ring binder. The two men spend about 20 minutes completing the observation and documentation process. After Jim returns to work, Randy spends another 5 minutes entering the checklist data into a spreadsheet into a computer.

On Monday morning, Randy shows employees at the weekly safety meeting the data from the previous week's safety observations. Randy tells the group he appreciates their efforts and to keep up the good work. He then discusses his concern about some employees not barricading work areas when called for. The employees as a group agree to try to achieve 100 percent proper barricading for the next four weeks.

On Friday afternoon, the behavioral safety steering committee holds its monthly meeting. The team of 10 includes Randy, employee safety representatives from each of several other areas in the plant, a representative from the safety department, and a representative from management. The committee reviews graphs that show the number of planned observations, the percent of employees conducting observations, and a bar chart that shows the practices that are cause for concern. After reviewing the data, the committee members decide they need to work on improving the use of fall protection. While it was not the most frequent cause for concern, the committee chose fall protection because of the potential severity of injuries that can result from falls. John, one of the committee members, agrees to chair a subcommittee to study the problem and develop an action plan for discussion at the next meeting. They decide that for this month they will simply share the data on fall protection with employees during safety meetings and discuss what can be done to improve this safety practice.

The steering committee also reviews several observations that members selected representing quality observations from the previous month. After selecting those that everyone agrees are "high-quality" observations, they make plans to review those observations at upcoming safety meetings and to provide a "Safety Champion" T-shirt to the employees who conducted those observations. The committee then adjourns and its members return to their work duties.

For yet another reporting period, not even a minor safety incident needs to be recorded or reported.

3 Value-Based Behavioral Safety Process

Culture is often defined as the practices common to a group of people. But it is much more than merely what people do; it is also the way they do things and the reasons they do them. In safety, we concern ourselves not just with the tasks that people do but also with how they do them. Also, we recently have become increasingly concerned about why they do the tasks in a certain way. When we talk about creating a safety culture, we are usually referring to creating an organizational environment in which people do their tasks safely and for the right reasons. The latter usually means employees perform tasks safely to prevent injury to themselves and others, not merely because of pressure from managers.

Unfortunately, most popular management literature suggests that changing organizational culture is difficult and can take years—not an encouraging prospect for a company that is struggling to improve a poor safety record. In addition, most of this literature is not particularly helpful for determining how to create a particular culture, much less a safer work environment. However, if culture also refers to the way employees go about their work, then we are talking about behavior, and a number of authors have described processes for improving safety through changes in on-the-job behavior (e.g., Geller, 1996; Krause et al., 1996; McSween, 1993a,b; Sulzer-Azaroff, 1982).

Such changes do not come automatically, however. In the following sections we consider why workplace cultures seem so much more resistant to change than individual behavior and why we are not always successful at implementing organizational change strategies.

3.1 A QUESTION OF BALANCE[1]

Figure 3.1 presents a simplified model of most organizations' culture and two paths that organizations use to achieve the desired results. It also shows the primary elements that must align to create a healthy culture within an organization. Thus, if an organization wishes to create an effective safety culture, it must create a vision or mission that describes its ideal, define values that clarify how employees will work together, and establish a process to achieve the desired results. Behavior is a particularly important element of this model because it is key to both processes and

---

[1] This section is based on Tosti (1993).

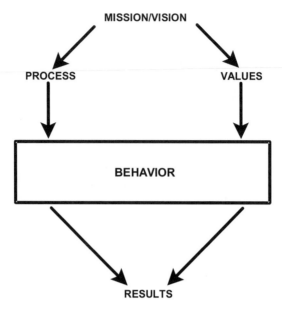

Figure 3.1. A simple model showing standard elements of organizational culture.

values. Good managers and healthy organizations establish a balanced emphasis on each of the elements in this diagram.

Unfortunately, many managers tend to focus on either process or results. Employees who report to these managers then reflect the manager's style. In addition, new managers tend to learn their management skills from those in the existing hierarchy. The organization itself may not focus its attention solely on process or on results, but because managers tend to hire and promote people like themselves, the emphasis may become and remain unbalanced. The result is that many employees and managers spend too much time on only one of these two elements.

Problems with an Overemphasis on Results. Many American managers manage their employees according to the results achieved by the employees. In many cases, we train managers to manage by results, then use antiquated systems of annual objectives and appraisals that maintain an unbalanced emphasis on results. Managers who enter the private sector from the military often have a strong results orientation, and industries that hire them, such as aerospace companies, defense contractors, and nuclear utilities, frequently have such an orientation. Companies that overemphasize results can be very successful, but they often achieve their success only through herculean efforts by top managers and key employees who often work extra-long hours. This orientation also frequently brings a decided personal cost. These companies typically have high levels of stress characterized by a high rate of employee burnout, high turnover, and higher than average rates of participation in employee assistance programs. Further, this approach often creates a perception among employees that "all management really cares about is the numbers," the reference being to budget figures, production records, efficiency targets, safety statistics, and the like.

In results-oriented companies, the tendency when the desired results are not achieved is to find someone to blame. The resulting fear and distrust often make improvement efforts difficult. Employees resist efforts to establish effective performance measures because they are skeptical about how managers will use those measures. They feel frustrated and believe the organization does not care about them. In fact, results-oriented organizations seldom systematically address how managers and employees should treat one another. The result is that the quality of personal relationships becomes a further barrier to any significant culture change.

In safety, *condition-of-employment* policies and incentive systems that unintentionally encourage employees to hide or overlook incidents often indicate an organization that overemphasizes results. Such policies implicitly threaten employees with the potential loss of their jobs, income, or promotional opportunities if they report an injury that results from a failure to comply with a company safety requirement. So, they may claim that the injury occurred at home or inaccurately report that the incident occurred in spite of full compliance with company procedure, saying something like "I was wearing my safety glasses, but this metal filing somehow got around them." They may respond similarly if a minor recordable injury jeopardizes their group's chances of getting a safety award and all the more so if the award is significant to group members.

Problems with an Overemphasis on Process. The relatively new focus for many American managers is on process. As quality improvement efforts grew more popular, managers learned to look more closely at the processes by which their organizations achieve results. Using such methods as clarifying requirements, standardizing procedures, and establishing measures for key steps in the work process, this approach tends to focus on the consistency of the work process that is often the basis for continuous improvement efforts.

Although such efforts work well for addressing process problems, many of the problems in today's organizations are not in process but in the relationships between employees and managers. Too often the tendency is either to create a new process or to fine-tune an existing one rather than deal with the root causes of relationship problems. Managers of nuclear utilities, for example, urge employees to identify potential nuclear safety problems. Their approach has been to create a process that encourages employees to report concerns and problems anonymously for independent investigation. The result, however, is often a growing bureaucracy that discourages personal responsibility and fails to address the root cause of the concern or problem. In many cases, the root cause is found in managerial practices that have destroyed the interpersonal relationship between managers and employees, thereby inadvertently creating distrust and discouraging employees from bringing matters to the attention of the appropriate line management.

Most of today's problem-solving tools are useful for addressing process issues, but they are often not adequate for addressing behavioral issues. Many of the tools assume that all problems are a function of the system while the efforts to resolve them typically add more and more complexity, bureaucracy, and paperwork. Such solutions increase costs and compete with other tasks that have more value to an organization. The bureaucracy in process-oriented organizations also creates the perception among employees that management only cares about paperwork. Employees who work in a process-oriented culture tend to attack problems and change processes without identifying or so much as considering problems that may be rooted in behavior and personal relationships.

Many companies have adopted an observation-based behavioral safety process to improve compliance with safety procedures, but simply implementing the basic elements of a behavioral safety process does not always work. Companies can encounter problems in getting employees to conduct needed observations, in getting employees to effectively provide feedback to other employees, and in employees filling out observation checklists informally without conducting actual observations. These problems may indicate an overemphasis on the reporting process and inadequate attention to the values required to implement safety improvements successfully.

Other General Management Problems. Conversely, individual managers may place too much emphasis on values, interpersonal relationships, or behavior without adequate focus on either process or results. Such an emphasis does not usually characterize an entire organization because the organization simply would not survive. It can occur, however, in the leadership of smaller groups such as teams or committees within an organization. Normally such groups have little conflict and everyone likes one another—but they are not very productive. Such groups often seem to take on a life of their own and go on forever, with frequent changes in direction but little actual progress toward problem resolution.

This problem is characteristic of safety committees that have difficulty in developing firm plans for achieving safety improvement. They tend to resist innovation and typically will not champion proven but different methodologies, such as behavioral safety. Rather, they continue to plan training for toolbox safety meetings and revise safety awards while implementing yet a further variety of programs that rely on the use of posters, slogans, and other approaches of dubious effectiveness.

3.2 ENSURE A CLEAR MISSION OR VISION STATEMENT

A basic tenet of current organizational theory emphasizes the importance of establishing a clear statement of an organization's mission or vision in order to establish a common purpose, which is the key element in a group becoming a team. The concept is reflected in the statement "Start with the end in mind." In many cases an organization already has formulated a safety mission or vision, or a mission statement in which safety is a primary element.

The basic purpose at this stage is to ensure that a safety design team has a clear mission. If an organization does not have a mission statement, the design team should work with upper management to create one. Once a working mission statement has been drafted, it needs to be reviewed by all employees in order to get their input and suggestions for its further enhancement.

3.3 TYPES OF MISSION AND VISION STATEMENTS

Mission statements generally fall into two categories. The first is a simple statement that clearly describes a desired state of affairs. This type of mission statement, in most cases a short phrase or sentence that is easy to remember, is one that many people refer to as a vision statement. The second type of mission statement is more elaborate and includes a statement of purpose. It also usually describes the products or services provided by the organization, often with details about aspects of the desired process for producing those products or services. Some organizations have a variety of vision or mission statements at different levels and for different groups.

Figure 3.2 provides examples of the two basic types of safety mission statements and considerations for their use.

Examples of Mission Statements	Considerations
Vision Statements Quality is job No.1! We will be world-class in safety. We will complete each day injury free.	a) Easy to remember. b) Provides basic level of guidance on the priority in daily responsibilities.
Purpose Statements The purpose of the XYZ Company is to ensure the successful completion of turnkey construction projects. We will provide construction and project management services in a manner consistent with the following principles.... Or, specific to safety: The purpose of the employee safety team is to design and implement a behavioral safety process in a way that • demonstrates our respect for fellow employees • ensures ongoing observations and feedback on safe practices • reinforces safe behavior on the job so that we remain number one in safety within our industry	a) Provides better understanding of the purpose of the business or team. b) Provides better guidance on qualitative aspects of the process. c) Is often posted publicly for ready reference.

Figure 3.2. Examples of different types of mission statements.

> The purpose of the employee safety team is to create a systematic process for managing on-the-job safety at the ABC Plant in a way that:
>
> - Enhances the meaningful involvement of all employees in safety activities within their areas
> - Establishes a formal safety observation process
> - Builds on and contributes to other quality and empowerment initiatives
> - Creates a positive work environment that builds pride in safety process and accomplishment
>
> so that the ABC Plant achieves significant improvements in safety as measured on the job by observation data and through OSHA-recordable rates.

Figure 3.3. A sample mission statement for a design team.

The process described below will help in developing a vision or purpose statement before beginning work on a behavioral safety process.

More complex vision or purpose statements often combine a statement of the expected results with value statements that describe how an organization or team wants to achieve those results. A common format has three elements:

Our purpose is to... (a statement of *what* the specific wants are that the organization wants to achieve).

To achieve our purpose we will... (several statements that clarify *how* the results will be achieved).

So that... (a statement about *why* these results are important).

Figure 3.3 provides an example of a safety design team's mission statement.

3.4 WHAT IS A VALUE?

As used here, a value is a statement or set of rules that prescribe the form of personal interactions preferred by a culture (such as the set of behaviors called "good manners"). Values serve as ground rules for interpersonal relationships within organizations.

Value statements may refer to either broad abstractions or specific practices. Figure 3.4 provides examples of each type of value statement.

Most organizations find it best to develop an initial set of value statements that are broad abstractions and then, as a separate step for communication and training, identify specific practices that will help employees understand the meaning of each statement.

Abstractions	*Specific Practices*
Concern for fellow employees	Resources committed to training and development
Open and honest communication	
	Diversity accepted and valued
Personal leadership	
	Recognition for achievements
Shared participation	
Teamwork and employee empowerment	
Continuous improvement	

Figure 3.4. Examples of the two types of value statements.

3.5 WHY CLARIFY VALUES?

Because value statements serve as ground rules for personal interactions, they are valuable in serving as behavioral standards for managers and employees in their interactions with one another regarding safety. Once an organization has developed statements of value and identified relevant practices that support those values, a basis has been established for providing training to ensure that both managers and employees understand how they need to interact with one another to make the behavioral safety process effective. In addition to their use in the initial training, these statements can also be a basis for ongoing evaluation and feedback to help ensure that everyone is interacting with one another in ways that will make the process successful and long-lasting.

By having employees discuss their values and practices that support established values, the organization's safety team can improve the initial success of the organization's behavioral safety process. Current research on moral decision making suggests that such discussions sharpen people's understanding of how the established values apply to different situations. Studies have reported that workers who have discussed how safety values apply in different situations are more likely to act in a manner consistent with those values when presented with an ethical dilemma (see Sherman, 1980). These studies suggest that employees who complete a well-designed, values-based training process will be more likely to honestly complete their observations and less likely to simply fill in the form based on informal observation. They will conduct the required observations and provide feedback not merely because of the procedural requirement but out of concern for their fellow employees.

3.6 USE A PROVEN PROCESS AND BUILD ON BASIC VALUES

Once the safety design team has agreed on its mission, it can begin work on the next two elements of our model: an organization's values and its process for ensuring a safe work environment. While it may want to clarify the organization's values before initiating the safety assessment, the design team will more commonly perform this task after its initial workshop. The advantage of this approach is in allowing the team to get a better understanding of the behavioral process and the kind of interactions required to make it successful before developing statements to clarify the values.

In Chapter 1 we described the elements that research shows to be effective in minimizing the risk of incidents, and forthcoming chapters will provide a detailed approach to implementing these elements. Again, simply implementing the elements of a behavioral safety process does not always work, often due to such problems as those described above (e.g., an overemphasis on process). The remainder of this book will discuss how to minimize such problems and thereby improve the integrity of an organization's safety improvement efforts by systematically addressing both the organization's behavioral safety process and its values.

3.7 CONCLUDING REMARKS ON VALUES

Safety professionals often acknowledge that an organization's culture is a key determinant of safety practices within the organization. Yet most efforts to improve safety are attempts to change processes, usually achieved through programs and procedures. A more complete approach to creating a safe culture includes both values and a proven behavioral process. Organizations must be able to systematically address how people interact with one another in order to maximize the benefit of new behavioral processes. A positive safety culture can demonstrably be achieved through a values-based approach to a behavioral safety process. Such an integrated approach leads to the kind of positive, action-oriented safety culture that some safety professionals and others usually only talk about.

4 Behavioral Safety Process

The following chart provides an overview of the stages in the implementation process. The first stage is discussed in Chapter 5, and subsequent chapters discuss later stages.

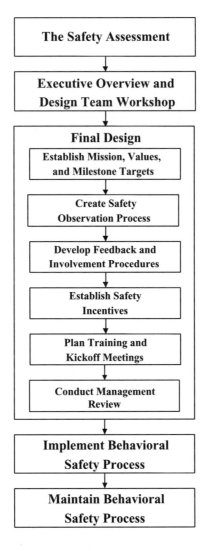

4.1 WHAT'S IN A NAME?

We call our approach to behavioral safety the values-based safety process. As you will design a process that meets the needs of your organization, your team may want to find a name that fits your organization's new process. Companies have called their safety improvement efforts the employee safety process (ESP), the safe acts process, the positive safety process, the continuous incident prevention process, and a variety of others. (See the discussion on the contest to name the behavioral safety process in Chapter 23 for suggestions on involving employees in selecting a name.)

A new name is most appropriate when you want to indicate that the new safety process is different from what was done in the past. However, you may choose not to give your behavioral safety process a name. If you are refining an existing process, you may want to emphasize just the enhancements. Giving your process a name may increase the chance that employees within your organization might initially react negatively to a change in direction and what they perceive as yet another new program.

4.2 TEAM-BASED PROCESS

The key to a successful behavioral safety process is getting the right teams together to plan the implementation. Figure 4.1 shows a typical team structure for the design and implementation phases of the behavioral safety process. Figure 4.2 clarifies who participates in each of these teams and some of their responsibilities.

The key to the success of this effort is to create the behavioral safety process through three successive revisions each addressing a different level of detail under the overall guidance of a management team or a health and safety committee:

1. During the initial assessment of safety factors, the design team develops its preliminary plans on the key elements of the process. These preliminary plans are simply a general framework for implementation, without all the logistical details that will come later.

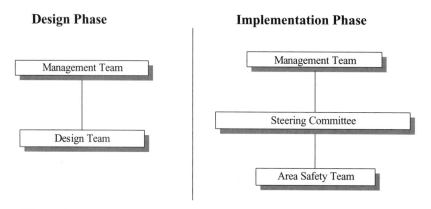

Figure 4.1. The team structure varies for the design and implementation phases.

Group	Who participates:	Deliverables
Management team, Central safety & health committee	Senior management (sometimes union leadership)	• Initial charter for design team • Approval and input into design team's plans • Active participation in support activities defined by the design team
Design team	Eight to 12 representative employees, a safety professional, a representative from engineering, and a representative from management	• Assessment of organization • Preliminary presentation to management • Implementation guides that includes procedures, forms, and training materials • Assessment of process trail and revised procedures, forms, and training materials
Steering committee	Same as design team	• Conduct and publish statistical analysis of observation data • Problem solve and ensure maintenance of behavioral safety process • Set improvement goals • Develop and carry out action plans
Area safety team	Supervisor, employee safety representative, and employees who normally meet for safety meetings	• Participate in observations • Set improvement goals • Implement action plans appropriate for their areas

Figure 4.2. Teams typically involved in implementing a behavioral safety process.

2. During the design phase, the design team (or teams) uses the preliminary plans to develop a detailed implementation plan, complete with forms, procedures, and training materials.
3. During implementation, the steering committees may tailor the basic design to the needs of their respective areas or locations.

As a rule, a single design team takes responsibility for the assessment and implementation planning. Sometimes during the assessment the design team will identify areas that are not represented and add team members representing those areas. Steering committees have primary responsibility for implementation. The management team periodically reviews and provides guidance throughout both the design and implementation phases of the process. Most organizations already have a safety committee structure with the equivalent of a management team and steering committees.

If such teams already exist, then the design team may define the responsibilities of existing teams for implementation and maintenance of the behavioral safety process.

The terminology used to identify different safety teams varies from organization to organization. The terms presented in Figure 4.1 will be used throughout book. In particular, the steering committee will refer to the group of employees that has final responsibility for implementation. This may be area or unit teams if you work in a large site or site safety teams if you have smaller locations that are geographically disperse. Depending on the size and complexity of the organization, you may also have some combination of these.

4.3 PROGRAMS VERSUS PROCESS

Currently, a popular management bandwagon is to create a *process*, not a *program*. A process is said to be never-ending, while a program has a discrete beginning and an end. Unfortunately life is not so simple. Certainly we need to create a *safety process* that is ongoing and relatively stable. The key elements of the behavioral safety process are regular observations, data review and problem solving, safety meetings, and safety recognition and celebrations. Most of this book is about creating a basic, ongoing process.

However, we also need safety programs. In an organization with a behavioral safety process, the steering committee implements safety programs to address areas of concern identified through the observation data. Such programs may include campaigns that provide novelty that keeps safety fresh. They help keep employees interested in different aspects of safety. Companies use ever-changing processes for marketing their products and they develop advertising programs that run for fixed periods of time. In the same way, an organization may use an ever-changing and evolving series of programs that support the basic process for managing safety.

The safety process provides ongoing measurement and evaluation. Under the umbrella of this ongoing safety process, you may want to initiate a series of programs that continue to use a set of basic concepts in new ways. The idea is to continually prompt consideration of safety factors. Constantly changing the programs helps maintain interest in your safety efforts by providing reviews of safety requirements in a variety of ways.

For example, you may run a program on back safety for three months or conduct a seatbelt campaign for employees and their families. The campaigns can include both behavioral interventions, such as training on proper lifting techniques, and efforts to address facilities issues that contribute to at-risk behaviors, such as reengineering lifting devices to make them easier to use. Safety campaigns and promotions are valuable but should not be continued until they become stale and lose their effectiveness. Rather, plan a fixed life cycle for each program; then end it and begin the next program or promotion. The key is that the targets are selected, and success measured, on the basis of the observation data. Chapter 23 provides an overview of a variety of other process elements and programs that may be important in support of your improvement efforts.

5 Safety Assessment

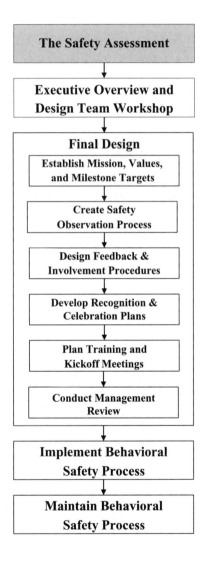

5.1 WHAT IS A SAFETY ASSESSMENT?

A safety assessment is a formal study of your organization's current level of safety performance and the practices that impact safety. It is not the same as a safety audit. A safety assessment examines the process the organization uses to manage safety. A typical safety audit, on the other hand, examines the extent to which employees are in or out of compliance with established safety procedures and environmental conditions are in or out of compliance with established requirements. Figure 5.1 contrasts the safety assessment and the traditional safety audit.

5.2 WHY CONDUCT A SAFETY ASSESSMENT?

Although you may be thoroughly familiar with your organization's safety improvement efforts, you should generally still conduct a formal safety assessment. Conducting the assessment serves two purposes:

1. An accurate and complete understanding of the organization's current safety efforts will result in better recommendations.
2. The interviews, presentations, and discussions resulting from the assessment will help you build support for initiating improvement efforts.

Conducting an assessment involves a team of employees in systematic data collection and careful analysis that should result in a common understanding of

Feature	Safety Assessment	Traditional Safety Audit
1. Observation of unsafe acts (behavioral sampling)	Yes	Yes
2. Observation of unsafe conditions	Sometimes	Yes
3. Summary of historical performance	Yes	No
4. Assessment of employees' knowledge of area safety procedures	Yes	No
5. Assessment of management practices affecting safety	Yes	No
6. Review of existing safety efforts (audits, safety meetings, awards, and special programs)	Yes	No
7. Final report identifies conditions that need to be corrected	Sometimes	Yes
8. Final report identifies unsafe behavior	Sometimes	Yes
9. Report includes recommendations for enhancing safety efforts	Yes	No

Figure 5.1. Key differences between a safety assessment and traditional safety audit.

your organization's safety efforts. The assessment provides an opportunity to check your understanding of safety improvement efforts in your organization through direct observation and independent verification. The assessment may prove or disprove your beliefs about the safety efforts in different areas and helps ensure that your perspective is not biased by experience with an unrepresentative part of the organization. It allows you to check the accuracy of your beliefs before you present your observations and recommendations to management or the health and safety committee. For example, you may suspect that personnel are not conducting safety meetings on the back shift, but you may not know that to be a fact. When two independent sources tell you that is the case or you show up at the scheduled meeting and find no one there, you will have a basis for your suspicion.

Using the assessment as a "reality check" will help ensure that your suggestions do not lose credibility because of inaccurate or unrepresentative information about the organization. Accurate observations regarding current efforts will give everyone greater confidence in your recommendations. If your audience agrees that you have described their current efforts accurately, they will be more likely to support your recommendations.

The assessment process also builds support for safety improvement efforts in other ways. Talking with employees about their involvement in safety efforts provides them with an opportunity for input. The discussions will increase everyone's awareness and understanding of your company's current safety practices and thereby provide a starting point for additional employee involvement later.

5.3 WHO SHOULD CONDUCT THE ASSESSMENT?

You will need to make an initial decision about who should be involved in the initial assessment and planning. The assessment can be either a team or an individual effort. Using a team requires greater coordination but will usually result in a better implementation plan and broader support for that plan. You should also consider use of outside consultants to either conduct the assessment or guide your assessment efforts. Outside experts are often useful in building management support for such efforts. Figure 5.2 outlines the options for who might be involved, the roles they may play, and possible advantages or disadvantages of each.

Figure 5.3 provides guidelines for selecting an outside resource to support your assessment, design, and implementation efforts. Be careful also to ask the questions given in the figure about the specific consultant who will be working with your organization, not just about the firm. As indicated in Figure 5.3, the ideal consultant should have 10 years of hands-on experience implementing behavioral safety programs, not including other types of safety experience. Consultants with less than 10 years experience should generally work with a project manager that has the appropriate level of experience.

Possible Participants	Potential Roles	Considerations
Individual manager or staff member	Conducts assessment and develops preliminary implementation plan	Simple logistics Time to conduct assessment Real or perceived biases can influence outcome Depending on experience, individual may need extensive training or guidance
Design team	Team conducts assessment	Builds broader support Better final product than individual alone Coordination is more complex May take longer to complete Team members often need extensive training or guidance May need support from outside consultant
	Team reviews interim assessment findings and final results	Builds broader support Better final product May take longer
Outside consultants	Conduct independent assessment	Adds objectivity Adds to short-term costs Often useful in building management support Consultant must have appropriate experience
	Provide assistance to design team or individual	Consultant can provide needed training and support Adds objectivity Adds to short-term costs Often useful in building management support Consultant must have appropriate experience

Figure 5.2. Considerations for deciding who should participate in an assessment effort.

5.4 OBJECTIVES OF THE ASSESSMENT

The assessment has several objectives:

1. Identify existing efforts and develop a plan that builds on these efforts.
2. Incorporate input from key personnel.
3. Identify high-risk areas and activities.
4. Identify training needs.
5. Identify potential design team members.
6. Build management support for implementation.

Questions to ask:	Ideal candidates will have:
1. What is the consultant's experience in implementing of behavioral safety improvement efforts?	Ten years of experience implementing behavioral safety improvement efforts Several customers that you can visit and talk with about their experience
2. What is the consultant's background in behavioral psychology?	Formal training in a program that specializes in behavior analysis Membership in the Association for Behavior Analysis Membership in the state chapter of the Association for Behavior Analysis (some state associations maintain a registry)
3. What industries does the consultant have experience with?	Experience with either your industry or a closely related industry that shares some of your problems and concerns
4. Will this consultant be available to maintain a relationship with your organization for the long term?	Consultant or project manager should have track record of long-term customer relationships
5. What level of onsite support should you expect?	Consultant should work as a member of your design team through all phases of the project, sharing responsibility for implementation Cost for this level of support should be clearly identified in the proposal
6. What travel costs will be involved?	Include an estimate of all travel costs when comparing proposals

Figure 5.3. Considerations for selecting a consultant to support your team in the design and implementation of a behavioral safety process.

Organizations often have different needs of an assessment. Sometimes the objective is simply to assess the readiness of the organization for behavioral safety and establish a team process for beginning one. For other organizations, a planning assessment is necessary so that both management and the design team have a preliminary plan for how the process will work. For some, management needs to know what it must do to build a foundation that will support an effective behavioral process.

The scope and depth of the assessment will vary extensively depending on the objectives. Some assessments are very formal and involved, requiring six to eight weeks to implement and the preparation of a formal report, while others may require only a few days and a 2-hour presentation to management. This chapter describes a full behavioral safety planning assessment. Many organizations will not require this extensive an assessment.

In some cases, organizations need more specialized assessments. An assessment of basic safety systems would check to ensure that the organization's basic safety management system is complete and functioning effectively. Other specialized assessments might examine organizational culture or leadership practices. These specialized assessments are beyond the scope of this text.

Identify existing efforts and develop a plan that builds on these efforts. Before developing your recommendations, you should identify the key elements of (1) your organization's current safety efforts, (2) what was done in the past, and (3) related improvement efforts. This history will enable you to develop and position your plan to build on existing and past efforts instead of creating an impression of a "new program of the month." During the assessment, you should also identify major initiatives to ensure that your safety improvement efforts align with them, especially those that have broad support. To maximize support for your behavioral safety process, you do not want to create an improvement effort that is at odds with other initiatives. For example, you will often need to coordinate safety and quality improvement efforts. A new safety process can easily be complementary to quality initiatives by exemplifying the philosophy of continuous improvement, using a team approach, analyzing data, and providing recognition.

Incorporate input from key personnel. As part of the assessment process, you should talk with people in each area of your organization. Such interchanges will provide them with an opportunity to have input into your plans. As mentioned, these discussions also begin to model the involvement that will continue later in the implementation phase.

Identify high-risk areas and activities. Another objective is to identify the high-risk areas and activities within your organization. You can do so via both interviews and a review of your organization's incident records. As you interview employees, you should ask them what they consider the most dangerous tasks in the area. This question will provide you with data on the work areas that employees perceive to be the most dangerous. Then compare this information with incident statistics to identify which work areas have the greatest risk and which jobs and tasks within those areas are highest risk. You can use this information to identify work areas that will be high priority and jobs and tasks that will receive special attention. It may also suggest the areas where you will want to pilot the behavioral safety process and identify practices you will want to be sure to include in your observation checklists.

Identify training needs. During your assessment observations, you may also identify skill deficits that are likely to have an impact on the success of your safety efforts. In the initial stages, the focus of your assessment will be on management and supervisory personnel rather than the employees. You should thereby learn how well supervisors conduct their safety meetings and how well they know the safety requirements in their areas. Later you will have to plan to address the deficiencies in these areas through training and coaching.

Identify potential design team members. One of the primary objectives of an assessment is often to determine who should participate on the design team. The best way to identify design team members is to take nominations from employees. Generally, the design team will be made up of 8 to 10 representative employees, a safety professional, one person from the engineering department, and a representative from management. During interviews, employees can provide the names of coworkers they would recommend for the design team. Ideal candidates are employees who have demonstrated an interest in safety, are well respected by their co-workers, and are assertive about asking questions and stating their opinions. The employees in each area who have the most nominations should be invited to participate on the design team.

Build management support for implementation. Finally, the most important objective of the assessment is to create management understanding and support for your implementation plan. Once you have completed your assessment, you should prepare a final report and schedule a presentation to management. Your report should summarize your findings and recommend a plan for enhancing your organization's safety efforts. The assessment report and presentation will provide you with an opportunity to ensure that (1) management understands what you are attempting to do and (2) it is willing to support the recommended improvements. This presentation is generally a decision point. If management agrees with your assessment and is willing to participate and support your recommendations, it will give approval to move ahead with your implementation steps.

5.5 OUTCOME OF THE ASSESSMENT

The primary outcome of the assessment may be a presentation or a formal report. Both would typically include the following:

- A summary of observations and findings
- The recommended team implementation process
- Suggested design team participants
- Preliminary plans for each stage of the project
- A detailed schedule and cost estimate

In your assessment report, you should include a summary of your observations and findings so that management understands the current situation that exists with respect to safety. The remainder of this chapter will guide your data collection and provide research-based standards for evaluating your organization's safety efforts. Your final report will then be primarily a matter of describing those safety efforts and making recommendations on how to implement changes that will move your organization toward closer alignment with the standards suggested by the current research.

The remainder of your final report should describe the implementation plan. During implementation, one of your goals is to ensure that you have broad ownership for the enhancements you are recommending. The only way to create such ownership is through involving others in the final design and implementation process. Therefore, as you conduct your assessment, you will plan a team structure to finalize and initiate the implementation plans. If the organization has an existing safety team structure, you will usually plan a team process that utilizes existing safety teams.

If your company does not already have safety teams, the assessment team will need to plan a safety team structure to ensure effective implementation and maintenance of the new safety process. Such team structures usually include two levels, one that provides management review and another that functions to develop detailed implementation plans based on the assessment report. The plan in the assessment report thus provides a preliminary design that the design team members use as a starting point for developing their detailed recommendations. The design team will generally develop additional details and implement the behavioral safety process, then move into a steering committee role once implementation is underway (see Fig. 3.1). During the assessment, you should identify both a team structure and a preliminary list of participants for both the design and implementation phases of the process.

The assessment report is the design document for implementation. The recommendations in the report serve as a preliminary design and provide a starting point for the design team. The design team can study the recommendations and decide whether to support, change, or discard the recommendations. Most often, they will revise the recommendations based on their experience and knowledge of the organization. They can then develop the additional procedural details required for implementation.

Finally, the assessment report may contain an estimate of cost and a preliminary implementation schedule. By including these items, you can give management a clearer picture of exactly how your organization might implement your recommendations and the potential costs involved.

5.6 SAFETY ASSESSMENT PROCESS

An assessment is somewhat analogous to completing a puzzle. First, you gather all the pieces. Then you complete the picture of (1) the organization's current safety performance and (2) what people within the organization are doing to achieve this level of safety. The assessment involves more than simply gathering data. It also involves analyzing the information on your current safety efforts, developing recommendations on potential improvements, and building support for those improvements.

How Do You Conduct a Safety Assessment?

The assessment generally consists of five steps, as presented in Table 5.1. Although these steps are presented in sequence for simplicity, in reality the first four steps often occur at the same time. For example, while you are conducting interviews,

TABLE 5.1. Steps in Conducting a Safety Assessment

Step Number	Activity
1	Review safety data.
2	Conduct interviews.
3	Observe safety meetings, safety audits, and safety practices in work areas.
4	Analyze information and develop an improvement plan.
5	Make the final report and presentation.

you may identify and arrange to visit a safety meeting well before completing the interviews.

Step 1: Review Safety Data

The first step is to review your organization's safety data, both the statistics and the actual incident reports. The incident data will help you understand whether the company has a significant safety problem and will be useful in planning how you will sell the importance of the proposed improvement efforts. Find out whether your organization is above or below the industry average. If your safety department does not have data about your industry, OSHA, the National Labor Board, or industry associations can usually provide such data.

If your rate of incidents is higher than the industry average, you should have no trouble building support for improvement efforts. Management will usually support a well-developed plan to move you to a leadership position in safety within your industry. If your rate of injuries is greater than the norm for your industry, safety improvements will produce significant dollar savings in insurance, workmen's compensation, and liability costs. Such savings will more than offset the cost of improvement efforts. If you are better than the industry average, you have a somewhat greater challenge because management's motivation for improvement may not be as strong.

The safety data should also help you identify groups with the highest and lowest incidence rates. Conduct the bulk of your interviews and observations with these groups to learn what factors account for their current levels of safety performance.

Also be sure to review individual incident reports to determine special hazards within the at-risk areas. What equipment and activities place employees at risk? When you begin your interviews, ask people to identify areas and tasks with the greatest risk and what employees can do to avoid incidents when working in these areas. Figure 5.4 provides a guide for completing the review of safety data. At this stage, you do not need to identify the behaviors that will prevent injuries. The design team should complete that task in developing the checklist as described in Chapter 9.

Activity:	Look for:	So that you can:
1. Review safety reports that summarize incident data	a) The rate of recordable injuries for the site or organization b) Areas with high injury rates	a) Compare with industry average to assess relative safety performance b) Interview personnel in high-risk areas
2. Review individual accident reports	Tasks and equipment involved in injuries	a) Prepare questions to ask about high-risk jobs during interviews b) Begin identifying specific tasks to include in the observation process
3. Collect data on compensation costs and, if possible, costs of property damage from accidents	Calculate an average cost per incident	Estimate cost savings that may result from reduced accident rates

Figure 5.4. Suggested activities for conducting a data review.

Step 2: Conduct Interviews

In your assessment, try to conduct interviews with people from a "diagonal slice" of the organization.

During interviews with managers, assess their interest in supporting additional enhancements to existing safety efforts. As mentioned, getting their support should be easy if safety has been a problem for the organization. If their areas have a good safety record, you may have a greater challenge. You may want to discuss how confident they are about how they achieved their current level of safety performance: Were they just lucky or do they really know what people are doing in their area to ensure safety? If managers have safety goals or objectives that specify improved numerical targets, they may be interested in developing a concrete plan for how they will achieve those objectives.

In conducting your discussion, be sure to get very precise answers. Any time an answer is not clear or specific, ask for more information or an example. You are looking for concrete observations, what people have directly seen or heard. When possible, you want to get concrete facts, not opinions, hearsay, or feelings about things. Try to validate information learned in the interviews through either corroboration in other interviews or direct observation in the next step.

As you are conducting your assessment, you need a standard of comparison to evaluate the information you collect. The checklist in Table 5.2 includes some of the questions you should try to answer through your interviews and observations. It also provides a recommended standard of comparison based on the research studies cited previously. This standard of comparison will give you a starting point for planning possible enhancements to current safety efforts.

TABLE 5.2. Safety Assessment Checklist

Questions to Ask	Standard of Comparison
1. Does the area have regular safety meetings? How often? How meaningful are they?	1. Managers or employees conduct short, weekly meetings that include relevant discussions of safety data, emergency procedures, near-miss incidents, etc.
2. Do supervisors and managers regularly talk to employees about safety? How often?	2. Besides safety meetings, supervisors talk with each employee about safety on the job at least once each week.
3. Does the work area have a formal process of safety audits or observations? How often? Who participates?	3. Supervisors and managers conduct daily or weekly observations and formally collect data on safety. Employees also conduct regular safety observations.
4. Do the observations focus on behavior or the environment? Do they identify problems or what people are doing right? What kinds of data come out of the process?	4. The observations focus on behavior and what people in the work area are doing right. The data yield the percentage of safe behaviors or a safety index.
5. Are the data graphed and reviewed with people in the work area?	5. Observation data are plotted on a graph that is posted in the area. Graphs and observation data summaries are discussed during safety meetings.
6. How are supervisors evaluated on their involvement with the system?	6. Supervisors are evaluated on how regularly safety observations are conducted in their areas. The evaluation includes how well they manage discussions in safety meetings and team reviews of safety data.
7. What kinds of safety goals are set in each work area?	7. The goals focus on process rather than incident reduction. Supervisors and employees set improvement targets for their work areas based on the safety observations or other data.
8. How does the area identify and respond to "near misses?"	8. The area has an active program to encourage employees to identify near-miss incidents. These are discussed in safety meetings and communicated to other shifts and areas. Employees may use such events as the basis for videotapes, slide shows, or other training materials.
9. What kind of safety training is provided?	9. Formal safety training is provided to new employees when they begin work in an area. The training includes a formal checklist for on-the-job orientation activities conducted by the supervisor. In addition, operators annually recertify in their units and go through a refresher course as necessary. Special training is scheduled in safety meetings to address issues identified in observations or by near misses. New supervisors, managers, and employees receive formal training in how to participate in the observation process.

TABLE 5.2 (*Continued*)

Questions to Ask	Standard of Comparison
10. What kind of safety award program is in place?	10. The safety award process provides recognition and celebrations of participation and success within the safety process. Awards are provided to teams that meet established goals and for individuals who report near misses and conduct all scheduled safety observations within the area, for example. Awards are not merely based on going some period without an incident.
11. How do managers and supervisors communicate the importance of safety as compared with production and cost?	11. Managers and supervisors make clear statements communicating the importance or safety. These statements are consistent in both group and individual meetings, regardless of the situation. Safety is the first agenda in every meeting. It is also a primary consideration in all personnel decisions (again emphasizing the process of managing safety, not simply the incident statistics).
12. How can employees identify safety problems in the area? What is done about problems so identified?	12. (a) The area maintains a safety suggestion system and employees are encouraged to identify safety concerns either through suggestions or in safety meetings. Suggestions may be signed or anonymous. The safety committee responds on area bulletin boards to all concerns. Signed safety suggestions are also one basis for safety awards and recognition. (b) Safety teams have a formal problem-solving process with clear, easy-to-use guidelines for assessing cultural issues and analyzing observation data, near misses, and incidents. In addition, they track significant issues to resolution and maintain a list of concerns with current status posted on the bulletin boards. Management works with maintenance to ensure timely responses to safety-related maintenance items.

Supplement the questions from the safety assessment checklist with additional questions you identify as you review the safety data and conduct the interviews. In addition to the items on the safety assessment checklist, here are some other questions that may be appropriate:

- What are the most hazardous tasks in the area?
- How quickly do safety problems get corrected?

- What are your current safety programs? What have you tried in the past? What other improvement or team efforts are currently underway?
- Do managers or supervisors directly or indirectly emphasize production over safety?
- What kinds of training have you received?
- Is equipment well maintained and safe to use?
- Who should participate in designing the safety improvement effort?
- Who should I talk with to learn more?

If you are conducting an assessment of a large organization, you may also want to conduct a survey of safety management practices. A survey allows you to get input from a large number of employees without the time required to conduct interviews. Perhaps more importantly, a survey can help you identify groups or pockets of employees for further study. By using survey data in combination with safety performance data, you can use interviews to gather information on safety management and leadership practices in both exemplary areas and areas that need improvement.

If you use a survey, you may want to conduct a small initial round of interviews to broaden your understanding of the kinds of issues you want to investigate through a written survey. Then, after you have analyzed the survey results, conduct more extensive interviews to gain additional understanding of the information that you gained from the survey.

Constructing a well-designed survey requires experience or training in survey construction. You must carefully plan the kind of information you want and what you will do with the information. Generally, items should ask about specific practices rather than attitudes or opinions. Consider developing categories of items that reflect your values and the practices that support or detract from those values. Obviously such items must be written in such a way that they are appropriate for everyone taking the survey. Figure 5.5 presents examples of possible survey categories and the kinds of practices that might be examined through a written survey.

Step 3: Observe Safety Meetings, Safety Audits, and Safety Practices in Work Areas

Try to visit several safety meetings and observe several typical safety audits in different areas. Some organizations will have different types of safety meetings and you should attend a sample of each. They may have a large monthly meeting and smaller, daily or weekly "toolbox" meetings. As you observe the safety meeting, you should attempt to answer the following questions:

- Are the meetings well run (is the agenda prepared in advance, do they stick to the agenda, is the discussion controlled, etc.)?
- Do employees participate in exercises or discussion?

Part I—Safety Leadership

Function	Practices
Direction	Area safety teams have clearly defined responsibilities.
	Employees discuss task hazards prior to starting to work.
Guidance	Area safety team receives feedback on adequacy of its efforts.
	Supervision provides feedback to employees on safe practices.
	Supervision provides feedback to employees on unsafe practices.
Motivation	Area safety teams receive recognition for their efforts.
	Employees receive recognition for their on-the-job safety.
	Employees are disciplined for unsafe practices.

Part II—Alignment of Values and Practices

Values	Practices
Mutual Support	Employees and supervisors provide assistance to other employees when appropriate.
	Supervisors provide time for employees to participate in safety process.
	Supervisors are evaluated on their support of the safety process.
	Management provides resources to support safety process.
Shared Participation	Employees are able to contribute their ideas and suggestions for safety improvements.
	Employees participate in discussions during safety meetings.
	All employees participate in safety process.
Trust and Respect	Employees respond nondefensively to feedback on safety.
	During safety team meetings, team members talk about others only in ways appropriate to their presence.
	Observers and supervisors ask for reasons when they observe employees working unsafely.
	Observers act as if they expect others to do their best.
	Supervisors discuss suggestions or concerns without being critical.

Figure 5.5. Sample safety culture survey categories and practices that might be examined through a written survey.

- Does the meeting include a review of current safety data, incidence rates, or safety process measures?
- Does the agenda include discussion of near-miss incidents?

Besides sitting in on safety meetings, you should also try to observe typical safety observations or audits. Safety staff, management, or other personnel often conduct these activities. A common problem is that they focus primarily on environmental conditions rather than on what people are doing in the area. As you observe the existing audit/observation process, you should try to confirm the information you learned during interviews regarding items 3 and 4 from the safety assessment checklist. Also, identify how observers respond to employees who are engaging in unsafe practices and how they respond to employees who are performing their jobs safely.

Activity:	Look for:	So that you can:
1. Attend safety meetings	a) Does the content of the meeting reflect the agenda? b) Do employees participate in exercises or discussion? c) Does the meeting include a review of current safety data or incidence rates? d) Does the agenda include discussion of "near-miss" accidents?	a) Identify additional questions for interviews b) Develop plans for improving the safety meetings
2. Observe existing audit or observation process	a) Who conducts audits or observations and how? b) What occurs when an employee is observed being unsafe? c) What is done with the audit or observation data?	a) Identify additional questions for interviews b) Develop plans for creating or enhancing observation process
3. Observe safety practices in work areas	Safe and unsafe work practices	Use as examples in draft safety checklists and training materials

Figure 5.6. Suggested activities for conducting observations during the assessment.

You may also wish to tour work areas and observe on-the-job safety. Use your experience to evaluate employees' use of personal protective equipment, housekeeping, and general safety. If you are unfamiliar with the work area, you may want to tour the area with an experienced employee or supervisor and ask them to identify practices that are safe or unsafe and where injuries have occurred in the past.

Figure 5.6 provides a guide for completing the observations during the assessment.

Step 4: Analyze Information and Develop an Improvement Plan

Once you have completed data gathering, analyze your information and develop an implementation plan. Analyze your information by comparing your organization's current safety management practices with those proven to be effective in the research studies described in the first chapter. Then develop a plan for implementing changes to your safety process that move your organization toward the model suggested by current research. The standards of comparison in the safety assessment checklist may provide additional suggestions for enhancing your organization's safety efforts. The outcome of this step is an outline for the final report and presentation that you will be preparing in the next step. You might prepare an outline that includes a single summary statement on the following:

- Significant observations of the organization's current safety efforts (e.g., "The area has one safety audit every year conducted by the safety department").

- Recommendations on how those efforts might be enhanced (e.g., "Create a process of weekly observations for collecting data on the areas compliance with safety procedures").
- Implementation steps (e.g., "Establish a design team comprised of front-line supervisors and representative employees to develop the observation checklist and procedures").

In developing your plan, decide who should participate in the design team that will be responsible for the final design and implementation process. Before deciding on design team members, make a preliminary decision about what areas are going to be involved. As stated earlier, the current research suggests all employees should participate in conducting observations. In Chapter 9, the section on how observers will be identified will discuss how to determine who should conduct observations.

Usually, before deciding who should be involved, you will have to decide what work areas are to participate. If you work in a large organization or at a large site, consider whether to pilot your behavioral process. A pilot is a good idea if your organization is likely to be particularly resistant to the idea. Most employees should view the pilot areas as representative of the organization, and it should usually have

- a high rate of incidents and
- management personnel willing to support the behavioral safety process.

You will also need to recommend whether observations should be conducted by managers and supervisors or employees or both. Many organizations are moving to a philosophy of greater employee involvement and self-managed teams. In organizations with a well-developed team process, employees should conduct safety observations. However, in some organizations safety remains primarily a responsibility of management and supervision. In such cases, the observation process should be part of the supervisor's job. Once the supervisors have a good understanding of the observation process, they then can begin to involve employees in the observations, often with the ultimate goal of involving all employees.

Once you have a preliminary idea of who should be responsible for the observations, you will have a better idea of who to involve in the design teams. As a rule, you need to involve the people who will initially do the observations.

The remainder of this book provides additional details on different options for addressing each element of the behavioral safety process and how to implement the process. Remember that your recommendations are preliminary plans and that the design teams will ultimately make changes and work out the many logistical details.

Step 5: Make the Final Report and Presentation

One of the primary goals of the assessment is to create management support. At the end of your assessment, your task is to explain to management how the organization can enhance its current safety efforts. Generally, you will want to prepare a short report that summarizes your observations and recommendations. However, do not assume that management will read your report; they may not. Your best strategy is to schedule a short presentation and discussion that allows you to summarize your assessment and discuss the recommendations. This approach will ensure that your management has a basic understanding of what you are trying to do. For simplicity, your report and presentation should follow the same outline. Figure 5.7 presents one possible outline for such a report and presentation.

Remember: This is a sales presentation. You are trying to sell your management on the value of implementing your recommendations. Before talking about the details of your recommendations, stress the realistic benefits that will result from a systematic implementation of the behavioral safety process:

- Employees who better understand safety procedures
- Greater compliance with safety procedures
- Exemplary injury rates
- Sustainable process

This presentation also provides an opportunity to get input from management team members and to incorporate their ideas as you move ahead.

You may want to schedule several individual meetings with key leaders before or after your group presentation. Use the individual meetings to explain your efforts and enlist support on an individual basis.

Assessment Report Outline

I. Introduction and Executive Summary

II. Assessment Methodology

III. Our Current Process:
    (use checklist items to determine best headings)
    A. Safety meetings
    B. Safety audits or observations
    C. Safety awards
    D. Management practices
    E. Data analysis

IV. Recommendations
    A. Team process
    B. Implementation steps
        1. Pinpointing safe acts
        2. Creating the safety observation process
        3. Feedback and involvement
        4. Safety awards
        5. Problem-solving & continuous improvement

Figure 5.7. Suggested outline for assessment report and presentation.

During your presentation, explain the kind of support and involvement required from management to make this effort successful. The behavioral safety process will usually require several forms of management support:

- Participation in the design review meetings and training
- Participation in weekly observations
- A review of safety data in existing meetings
- Allowing employees time to participate

The goal of these discussions with management is to get a "go-ahead" decision. This decision should include agreement to actively participate and support the design team's improvement efforts.

6 Management Overview and Initial Workshops

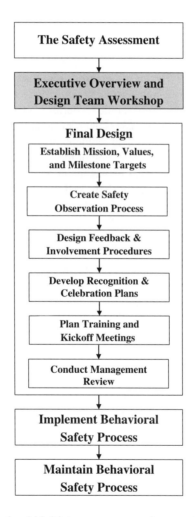

To ensure success, you should initiate your process in a carefully planned and well-organized manner that ensures a high level of understanding of the changes you propose. You will have at least three groups to consider: managers and supervisors, those who will be participating in the design team, and other employees. You will need a strategy that combines communication and training. In the management overview, you provide managers and supervisors with an outline of your plans for enhancing their current efforts. The overview will tell them what the design team will be doing and what they can expect as the process develops. By providing a behavioral safety workshop, you will give design team members an understanding of their tasks and the basic set of the skills they will need for planning activities. You will also establish a group of knowledgeable employees in the work area who can provide information to others about the behavioral safety process. Later, the design team will have to plan additional communication and kickoff meetings with employees.

6.1 WHAT IS THE MANAGEMENT OVERVIEW?

A management overview is an orientation for management and supervisory personnel that introduces the enhancements you plan to make to the organization's safety improvement efforts. You should generally invite management personnel who were not present at your assessment presentation, especially those from the work areas targeted for initial implementation. This meeting may have the same basic agenda as the final presentation of the assessment results except that you are now explaining plans that have management approval.

Despite management's approval, you still need to sell the value of your recommendations. In particular, you should ensure that your audience understands the basic elements of the behavioral safety process and how they can have input into its final design. The goal is to ensure that managers and supervisors in the area know about the proposed changes. This understanding will remove fear of the unknown and help build their support for the new process. Their support will help you overcome resistance to the changes required in a new safety improvement effort. You should also let them know how the process will evolve and when they can expect to learn more about the design team's plans.

If your organization involves shift work, you will probably need to hold several such meetings to ensure the participation of supervisory personnel from all shifts.

Depending on the size of your organization, you may want to provide a series of kickoff meetings for all employees, instead of just managers. These may be short overviews presented as part of regular safety meetings or a separate series of meeting. The purpose of the meetings remains basically the same as for the management overview described above.

6.2 WHAT ARE THE OBJECTIVES OF THE MANAGEMENT OVERVIEW?

These meetings have several objectives:

- To provide an overview of the elements of the behavioral safety process and why it is being implemented

- To ensure that line management knows how to support the implementation process at this stage
- To let people know how to influence the process

6.3 WHAT IS THE DESIGN TEAM WORKSHOP?

The initial workshop is similar to the management overview except that it usually includes formal training for those who will participate in the design team (refer to Fig. 3.1 for the typical team process). If your design team involves primarily supervisory personnel, you may only need one meeting. This initial workshop will generally include all of the people who will be participating in the design meetings. You may want to include a variety of other personnel. For example, you might include volunteers from each shift and area involved in the implementation. You may choose to include union leadership or people from work areas that may develop a similar effort in the future. In particular, you should consider including informal leaders who will help support your effort later if they understand what you are trying to do. Again, one of your goals is to let people know how they can have input in the design process and where to go if they have questions.

6.4 WHAT ARE THE OBJECTIVES OF THE DESIGN TEAM WORKSHOP?

This workshop has several objectives:

- To provide the necessary skills to those who will participate in the design process
- To build better understanding and support from key managers and employees
- To get initial input from participants as a basis for later development

6.5 WHAT IS THE AGENDA?

The behavioral safety workshop is typically three days depending on the level of skill you want to provide to the design team and the amount of detail you want to provide to other participants. Figure 6.1 provides a typical agenda for a typical behavioral safety workshop. Formal exercises or small group discussions follow each agenda item. These exercises help ensure that the participants understand what they will do in the implementation process. The agenda parallels the organization of this book, which might be used as a supplement to the workshop. It might also include a module on developing values and identifying practices aligned with those values along with additional exercises on each of the key elements of the behavioral safety process.

> **Design Team Workshop**
>
> **Part 1: Basics**
> - Safety basics
> - Analysis of traditional safety programs
> - Pinpointing behavior
> - Our values
>
> **Part 2: Implementation Overview**
> - Pinpoint safe acts
> - Safety observations
> - Feedback and involvement
> - Recognition programs
> - Maintaining the behavioral process
>
> **Part 3: Behavior Analysis**
> - ABC analysis of safety
> - Behavioral incident investigation

Figure 6.1. A typical agenda for a behavioral safety process kickoff workshop.

This meeting should serve as the kickoff meeting for the design team. You may want to establish ground rules for later meetings, and even conduct an exercise that enables team members to develop a preliminary purpose statement in preparation for your first design team meeting.

7 Final Design

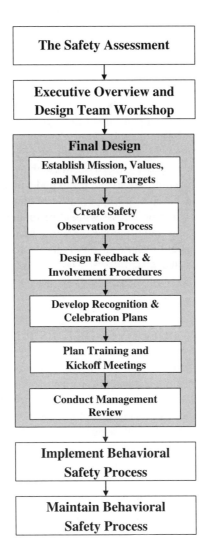

Here you will be working with the design team to carry out the next phase of your recommendations: the development of a detailed implementation plan for the new safety process within your organization. If a team conducted your assessment, some of the assessment team may also be participating with you on the design team. The recommendations from your assessment along with initial ideas captured during the design team workshop will serve as a preliminary design and provide a starting point for the design team's efforts. This way you will not walk into a design team meeting and begin to brainstorm what the team members think needs to be done. Rather, you start by presenting the assessment recommendations and the rationale for those recommendations along with the team's input from the design team workshop. After this presentation, you change hats and become a participant in the meeting. As a participant, you work with the other team members to finalize the implementation plans.

Design team members may occasionally decide to proceed in a different way than you recommended on the basis of the assessment. As a team member, you have input into the team's decision, but you cannot force your views. If the team wants to do something that differs significantly from the plan originally approved by management, you will need to go back to the management team and reach agreement on the team's purpose. Otherwise, trust the team process. If you do not agree with a course of action, say so and try to find a mutually agreeable course of action. You will often be able to support a team member's ideas while you continue to champion progress on the assessment recommendations.

Over the long term, the safety process should be responsive to data and experience. The group should ensure an observation process that gets modified on the basis of incidents and near misses. The important thing is to create an adaptive system that responds to experience in ways that will help employees prevent future incidents and injuries.

7.1 WHAT ARE THE OBJECTIVES OF THIS PHASE?

In this phase, your goal is to work with the design team to complete detailed plans for enhancing your organization's process for managing safety in the work areas. Completion of the implementation plans usually means developing additional details for carrying out the assessment recommendations. The specific objectives include creating or enhancing the following:

- A safety observation process that provides a regular measure of on-the-job safety
- Feedback procedures for communicating and reviewing observation data
- Safety awards tied to on-the-job safety improvement
- An involvement process that ensures ongoing involvement of all employees

The remainder of this book will describe a design and implementation process to be followed as though you are initiating a completely new safety initiative. You should be able to adapt these basic procedures if you are simply refining your existing efforts.

7.2 DESIGN TEAM PROCESS

A key element of the design team's job is to clarify the responsibilities of steering committees both during and after implementation. You will need to establish such safety teams if they do not already exist. In large organizations, the design team's plans should clarify the responsibilities of local steering committees. These may be site safety teams if your organization has relatively small locations and they may report to a divisional or centralized company safety team. They may be area or unit safety teams within a larger plant and report to a site steering committee or central safety team (see Fig. 3.1). The design team(s) will also need to plan the responsibilities of management for supporting the new process. The design team should operationally define the responsibilities of these groups for both their initial participation and their long-term support of the safety process. The design team may additionally need to plan its own responsibilities during a trial, or pilot, process and its responsibilities as a steering committee if it plans to continue in this capacity after implementation.

The details of your implementation plan will vary depending on the size and complexity of your organization. Figure 7.1 provides guidelines for planning the implementation structure and process. For large organizations or organizations that are spread out geographically, you will usually want to conduct day-long planning meetings in order to complete your planning prior to beginning implementation. We have found that two-day planning meetings are most effective. You will need to plan a logical implementation sequence, then bring steering committee members to a central location for the first phase and provide a single workshop that covers all elements of the process. Next, you can provide similar implementation workshops at central locations for the steering committees in the later phases. You should

If your organization is	Then
Small (300 people or less) at a single location	The design team and steering committee will be the same, often using periodic planning meetings
Spread across multiple locations	The design team should conduct two-day marathon planning meetings to plan the process and develop materials, then implement through steering committees
A large, single location	The design team may use either marathon or periodic planning meetings, then implement through steering committees

Figure 7.1. Guidelines for planning the implementation structure and process.

evaluate training and implementation efforts at each stage so that you can refine the process throughout implementation.

If your organization is resistant to change, you should consider doing a pilot in work areas where you are confident you can achieve success. The ideal pilot areas should (1) be viewed as representative of the other work areas of the organization, (2) have managers and personnel who are likely to be supportive of the process, and (3) definitely need improvement indicated, for example, by incidence rates.

The implementation process described in this book assumes you have a design team that will complete multiple two-day marathon planning meetings, then implement the process through steering committees. The same process can be adapted for simple organizations where the design team will become the steering committee. The periodic planning meetings are discussed at each of the primary stages when the process differs from the two-day planning process.

The time required for implementation will be a function of the size of your organization and whether your planning and implementation are concurrent or sequential. As a general rule, conducting several day-long planning meetings is preferred for larger organizations because it reduces the length of time required to complete the planning and implementation process. Figure 7.2 provides an overview of these options along with guidelines on the time required for implementation in a relatively simple organization or a single component of a more complex organization.

Whether you choose a periodic or a marathon planning process, more complex organizations, organizations that are resistant to change, and organizations that lack strong and active management support are more likely to need outside assistance. An experienced outside consultant can help you avoid mistakes and build the credibility of your improvement efforts. Such an outside resource also is more important in the marathon planning process because the schedules are usually more compressed.

Design team meetings options:	Considerations:
Periodic planning meetings (2-hour weekly or biweekly meetings)	Allows implementation of observation process before planning other elements of the process, often as soon as 2 months after beginning planning Takes longer for implementation, usually 8–12 months for planning and implementation in a small- to medium-size facility
Marathon planning meetings (2-day, off-site meetings)	Minimizes length of time required for planning and implementation, usually 4–8 months for initial implementation Facilitates design team participation from multiple sites with minimal travel costs

Figure 7.2. Considerations for selecting periodic or marathon planning meetings.

7.3 ROLE OF THE SITE MANAGEMENT TEAM

The site management team has several functions during this phase. Its first responsibility is one of project management, especially if the design team is using a periodic planning process. It should provide the design team with an initial goal or purpose statement, then ensure that the design team develops its own initial goal or purpose statement in a manner that is consistent with the steering committee's intent.

If the design team uses periodic planning meetings, the site management team should also review the design team's schedule and agree on the milestone reviews. One of the difficult aspects of implementing this kind of change effort is establishing a sense of urgency. A schedule will assist the design team's leader in maintaining progress. (See Appendix A for sample schedules.) The design team leader should generally provide regular updates to the site management team on the design team's progress.

The site management team is also responsible for changing organizational policies and procedures to better support the behavioral safety process. In particular, it should deal with issues that are beyond the span of control of the design team. These might include making changes to company incentive programs, performance appraisal, and compensation systems in ways that support the new safety process. The site management team may also work closely with the design team on defining management's role to ensure maintenance of the safety process. The assessment should identify and include recommendations for addressing these issues.

Finally, the management team will be responsible for approving and sanctioning the design team's recommendations. Once the management team approves the team's plan, the design team proceeds with implementation. The management team's approval should include commitment of resources, including management participation and funding for various elements of the process, such as time for meetings, money for training, travel costs, safety awards, and related activities.

What Are the Steps in This Phase?

Generally, design and implementation efforts will follow the six-step process given in Table 7.1.

TABLE 7.1. Typical Steps in Designing Behavioral Safety Process

Step Number	Activity
1	Establish a mission statement, values, and milestone schedule.
2	Create the safety observation process.
3	Develop feedback and involvement procedures.
4	Develop recognition and celebration plans.
5	Plan training and kickoff meetings.
6	Conduct a management review.

For simplicity, these steps will be presented as a sequential process. Although this description may not always apply, it captures the critical tasks and serves as a guide for your design efforts. If your design team is conducting periodic planning, some of these steps may occur almost at the same time or in a different order. For example, you may conduct the area kickoff meeting (step 5) and initiate the observation process before addressing the planning involved in steps 3 and 4. If this is your approach, be sure to review each element with management prior to its implementation.

Also, especially in periodic planning meetings, individual group members will frequently champion one or more special programs (see the discussion on providing for a variety of ongoing support programs in Chapter 23) while you are planning elements of the other steps. The team should encourage such involvement by supporting such initiatives. Seeing their ideas come to life will help maintain participants' interest in working on the safety team.

8 Step 1: Establishing Mission, Values, and Milestone Targets

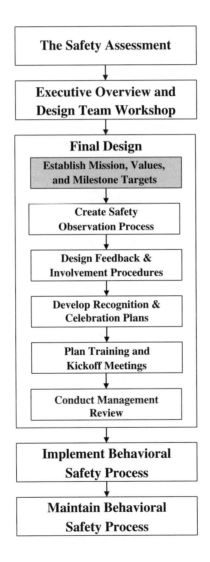

STEP 1: ESTABLISHING MISSION, VALUES, AND MILESTONE TARGETS

Before starting to finalize the design of the safety process, work with the design team to develop a goal or purpose statement. In some cases, you may have a charter of the initial statement of purpose from the management steering committee or a draft produced by participants in the initial workshop. If you do not already have one, draft a preliminary purpose statement before the first design team meeting. Or you may choose to develop the mission or purpose statement as a group exercise during the meeting (review discussion of mission statements in Chapter 3). Also, prepare a sample schedule based on your best guess about the activities and the amount of time required for each. Then, in the first meeting, allow the team to discuss, modify, and approve both the purpose statement and the schedule for presentation to the management team. Figure 8.1 presents a sample goal or purpose statement.

To: Implement improvements to our process of managing safety on the job

In a way that:

    Provides regular observations and feedback for safety practices in our work areas,

    Establishes increased employee involvement in observations and safety meetings, and

    Provides safety awards for achieving high levels of compliance with our safety practices,

So that: We achieve a safer work place as measured by both our safety observations and our rate of OSHA-recordable incidents.

Figure 8.1. A sample purpose statement for a design team.

8.1 CLARIFYING VALUES: A STRUCTURED APPROACH

After five years of using an unstructured approach to assist organizations with the task of defining their values, we developed a more structured process using common themes that kept recurring as we worked with different organizations. Outlined below, this structured approach requires less time and provides a set of value statements and corresponding practices found to be just as effective in guiding the development and assessment of the new behavioral safety process.

An unstructured approach may still be appropriate for organizations that want to develop a very personalized set of value statements with corresponding practices. Appendix C provides the guidelines for our initial approach.

The design team that will be planning the new process for the organization typically undertakes this exercise. If the design team is large, breaking it down into smaller groups may best accomplish the first three of the following steps and assist your organization in creating a value-based safety improvement process:

Step 1: Brainstorm actions likely to impact the process.
Step 2: Pinpoint those practices.
Step 3: Sort these practices into "value" categories.
Step 4: Use values in designing your safety process.
Step 5: Discuss values during kickoff meetings and training.
Step 6: Use values as criteria for evaluation.

8.2 STEP 1: BRAINSTORM ACTIONS LIKELY TO IMPACT THE PROCESS

Brainstorm a list of positive and negative actions that would either support or damage your efforts to implement a new safety improvement initiative. This needs to be a true brainstorming effort. Simply capture everyone's ideas on a flip chart. Write them all down on the chart without editing what they say. Do not try to make the items more specific or identify pinpointed behaviors. At this stage, you will typically get very general statements such as "Need visible management support" on the positive list and "Lack of communication" on the negative list.

8.3 STEP 2: PINPOINT THOSE PRACTICES

Pinpoint all of the items on your lists. (For more detailed instructions on pinpointing, refer to Chapter 21.) Restate each of the positive items on your list to describe a specific activity or behavior. Also restate each of the negative items in the positive; in other words, restate items that would damage safety efforts as statements of actions that would have a positive impact on the process. To use the examples given above, "Need visible management support" might be restated as "Get upper management to introduce the safety process and explain why it is important." "Lack of communication" might first be stated in the positive as "Develop good communication," then further stated as "Keep employees informed about planning and the implementation progress." Each pinpoint should be a description of a specific behavior that demonstrates support for the process. In addition, each needs to be worded in language that is meaningful to the employees who will see them.

8.4 STEP 3: SORT THESE PRACTICES INTO "VALUE" CATEGORIES

Most design teams identify activities that can be readily categorized into the following value categories:

- Concern for the well-being of others
- Open and honest communication
- Personal leadership

- Teamwork and employee involvement
- Continuous improvement

Have each group sort their pinpointed practices into the value categories. If you have activities or practices that do not correspond to one of these values, you may need to modify the categories or perhaps add one or two additional value statements.

After listing all the group's responses, review each list to ensure that each practice is listed under the value that it best supports. Also look for closely related practices that might be combined. The ideal number is between three and seven practices for each value. More than seven makes the list of values and associated practices too long and complex to provide practical guidance to managers and employees.

At this stage you should combine the lists produced by each of the small groups. Revise and combine similar items. Once you have condensed the list as much as possible, have each group select the three to five practices they consider most important in each category. Next, create a master list of values and practices by combining the lists prepared by each group. Again, as a group, condense and combine items as much as possible, then once again select the three to five practices in each value category that will be most important to the success of your process. Continue this process until you have achieved a consensus on the three to five practices that are most critical to the success of the process in representing each value.

Once you have identified your initial set of values and related practices, go back over the list to see if you can identify any other practices that relate to each value statement that should be considered.

After getting an appropriate number of practices for each value statement, do a final edit to ensure the specificity and clarity of your practices. Also try to ensure that your practices are mutually exclusive. If two of your practices are similar or potentially overlapping, either combine them or rewrite them to make them more distinct.

Figure 8.2 presents examples of critical practices related to the value statements identified above.

8.5 STEP 4: USE VALUES IN DESIGNING YOUR SAFETY PROCESS

These values and critical practices should guide you in the many decisions that you will need to make in designing your process as outlined in the sections that follow. For example, some of your critical practices will specify roles and responsibilities for those who will be involved in the process, including observers, steering committee members, supervisors, managers, and support staff. These practices may include roles that you want to define formally as procedures or responsibilities within your behavioral process. Other critical practices may specify behaviors that you want to use to provide guidelines on how people interact with one another within your new process.

Value Statements	Critical Practices
Concern for fellow employees	Provides feedback to others to help them identify and minimize risks
	Identifies potential risks and actions needed to eliminate or reduce them
Continuous improvement	Identifies and documents problems and opportunities for improvement
	Ensures that improvements are properly executed
	Promotes improvements to the behavioral safety process
Personal leadership	Trains/coaches others in safety processes
	Sets examples for others
	Makes time for participation in safety activities
	Provides recognition for safety practices and accomplishments
Open and honest communication	Listens well and responds appropriately to others
	Follows through on verbal commitments
	Communicates safety-related information promptly
Teamwork and employee empowerment	Participates and contributes to team effectiveness
	Solicits ideas and suggestions from others
	Strives for consensus

Figure 8.2. Examples of value statements and corresponding critical practices.

8.6 STEP 5: DISCUSS VALUES DURING KICKOFF MEETINGS AND TRAINING

Be sure to discuss your values and critical practices when you introduce your behavioral safety process to employees. Kickoff meetings are ideal times to communicate your expectations about how employees are to participate in your new safety process. Discussing your values at these meetings will communicate your expectations from the start as to who will participate and why. When discussing "open and honest communication," for example, you can talk about the importance of (1) having an open and honest discussion about what is observed and (2) recording accurate information on the observation forms.

The purpose of these discussions is to clarify the organization's expectations and sharpen the employees' understanding of each value statement. You will usually not have to conduct extensive skill development exercises because most employees will already have an intuitive understanding of what these statements mean. The purpose of the training is to clarify situations that are appropriate for applying their existing skills and knowledge. In some cases you may need to provide additional training to help employees develop or refine the communication and leadership skills necessary for a successful safety process. Other approaches to this training might utilize case studies for analysis and discussion. The goal of these exercises is to get participants to consider how they will respond when faced with situations that they are likely to encounter as you implement your behavioral safety process.

8.7 STEP 6: USE VALUES AS CRITERIA FOR EVALUATION

In your safety process, values have two additional uses:

1. They provide a basis for safety teams to evaluate the quality of the new safety process once it is up and running.
2. They provide a basis for evaluating individual performance within an organization's performance appraisal process.

When your employees complete their initial observer training, your organization's newly established values should serve as ground rules for their interactions with one another. The statements of values and practices provide a basis for reciprocal feedback and for evaluating actions and potential actions. The question "Did this particular practice support our values?" should be a standard part of evaluating past actions, and the question "Does this decision support our values?" should be a standard part of decisions regarding future actions. In short, the values can serve as guidelines for the kinds of behavior expected within the organization and thereby clarify behavior that members of the organization will support.

Your steering committee can use these values and practices as a basis for evaluating the quality and integrity of its safety improvement efforts. Chapter 15, on maintaining the behavioral safety process, presents additional details on using values as part of the assessment process.

8.8 ESTABLISH A MILESTONE SCHEDULE

Your team will also need to develop a milestone schedule for completing each stage of planning and implementation, especially if you are conducting periodic planning meetings. If you are using the two-day planning meetings, you may simply schedule a presentation to your management team during your final meeting. Otherwise, the schedule should include milestones that indicate when the team expects to review its plans for each implementation step with the management team. These design reviews not only allow an opportunity for management to have input into your plans but also ensure that management is informed and willing to support those plans. As a minimum, the design team should plan to meet with the management team to review the completed plans for the following:

- Observation process
- Feedback and involvement procedures
- Safety incentive program
- Kickoff meetings and observer training

After implementing the basic elements of the behavioral safety process, the design team may need to schedule additional steering committee presentations to seek approval and support for various special programs (described in Chapter 23).

You should plan the equivalent of six full days for planning the implementation of the behavioral safety process. If your team is conducting two-day planning meetings off-site, schedule three two-day meetings to complete your detailed planning. Your team may complete its tasks more quickly, but this is a rough guideline for your initial planning. As stated above, with the two-day planning meetings, you will probably only need a single meeting with the management team to review your final plans for implementation. If your team is conducting periodic planning meetings, then you may want to schedule management presentations after critical milestones. Then your team can implement each component as it begins to plan the other elements of the process. Regardless of your approach, you will need to coordinate your schedule with management to ensure that they are available to participate in the design review(s).

Appendix A provides a sample implementation schedule for both the biweekly and two-day meeting approaches to planning.

9 Step 2: Creating the Safety Observation Process

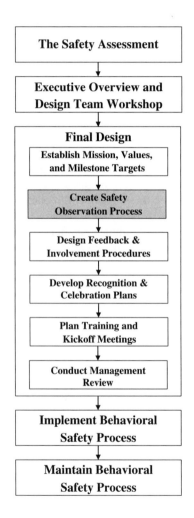

Note: The approach discussed in this chapter is based on Sulzer-Azaroff and Fellner (1984). We encourage you to read the original article, which includes an excellent example of a well-designed safety observation checklist from a paper mill.

STEP 2: CREATING THE SAFETY OBSERVATION PROCESS

The foundation for maximizing the effectiveness of your safety process is the regular observation of safe practices. Early studies suggested that monitoring work activities was the single most important activity a manager could engage in to ensure good performance (Chhokar and Wallin, 1984; Fellner and Sulzer-Azaroff, 1984; Komaki, 1986; Komaki et al., 1986, 1989). More recent studies show how important observations are in improving the safety performance of observers (see Chapter 27). An effective observation process offers the following advantages:

- Improves the safety practices of observers
- Results in better hazards recognition
- Provides feedback on the effectiveness of the safety process
- Establishes a baseline for setting improvement targets
- Provides practice in observing and discussing safety
- Adds social consequences for safe work practices and provides a basis for additional forms of employee recognition

As discussed previously, one of the benefits of an observation process is that employees who conduct observations learn to work more safely (see the discussion of the observer effect in Chapter 27). Observers begin to perform the behaviors on the checklist much more consistently, surprisingly even when they have not been observed or received feedback on their own safety practices. In addition, by conducting safety observations, employees are learning to identify hazards in their work areas, which may, at least in part, explain why they begin to work more safely.

The observation process also provides a measure of the success of your safety management process. By collecting data on safe practices, you have an *in-process* measure of safety. This use of observation data is in contrast with the traditional use of incidence rate data, which is an outcome measure of safety. An observation process provides an opportunity to work proactively with employees on their work practices before someone gets hurt. It also enables you to evaluate the effectiveness of specific safety improvement efforts. If you conduct a campaign on back safety, for example, the observation process can provide data about whether employees are lifting safely. In addition, an observation process can identify possible problem areas regarding work activities. In this case, the observation data might indicate that you needed a program on back safety before a worker injured his or her back.

This approach to safety parallels the quality improvement efforts of many organizations. One of the key features of such efforts is a focus on the process rather than simply on outcomes. In safety, paying attention to the process means paying attention to safe practices.

The observation process directs attention to and arranges feedback on how employees are doing their job. Several studies suggest the importance of verbal feedback in establishing and maintaining safe practices on the job (e.g., Komaki et al., 1980). The observation process arranges for regular opportunities for such feedback and helps ensure that it occurs.

The observation process also provides a source for data for employees to use in identifying practices that need improvement. In several of the studies discussed earlier, employees used observation data as the basis for setting improvement targets (cf. Sulzer-Azaroff et al., 1990). The data enable employees to set improvement goals based on what they had been doing and what they thought they could do better. They can then develop action plans to achieve those goals.

One reason for establishing a behavioral safety process is to help employees within the organization get into the habit of talking to one another about safety. Our ideal organization establishes a culture where each employee is his brother's (or her sister's) keeper. Operationally, this practice might mean that everyone is paying attention to how co-workers are doing their jobs and regularly talking with them about the safety of their work practices. In reality, using the typical informal approach, it is difficult to get supervisors and managers consistently to attend to safety compliance. A behavioral safety process ensures that everyone, managers and employees alike, knows how to observe and provide feedback on safety. It then ensures that everyone practices observing and talking to one another about on-the-job safety.

To summarize, the observation and feedback process is probably the most important element of the behavioral safety process. The design team must design a realistic observation process, complete with appropriate forms and guidelines, on the logistics of conducting observations within the organization.

9.1 HOW DO YOU CREATE THE OBSERVATION PROCESS?

The steps in the implementation of the observation process are outlined in Table 9.1.

TABLE 9.1. Implementation of Observation Process

Task Number	Activity
1	Analyze past incidents and injuries.
2	Develop a list of critical safe practices.
3	Draft and revise checklists.
4	Develop the observation procedure.
5	Trial run the observation checklist and process.

9.2 ANALYZE PAST INCIDENTS AND INJURIES

The first task in developing an observation procedure is to analyze the incidents and injuries within your organization. Typically you will want to review every incident that has occurred within the last three to five years, then determine the practices that would have prevented the injuries from occurring. During this analysis you have five objectives that will be important as you develop your observation process:

1. Identify critical behaviors for observation checklist(s) based on both the frequency of their occurrence and their potential severity.
2. Identify whether severe injuries most likely occur during routine or nonroutine operations.
3. Identify department-specific behaviors for use on department-specific checklists (optional, depending on assessment results).
4. Identify specific examples to include in definitions.
5. Identify the time of day and days of the week that injuries are most likely to occur.

Your design team members should review past incidents and near-miss records to identify unsafe practices that caused or nearly caused incidents in the past. These incidents should be fundamental in constructing your list. Prior to beginning your analysis, you should sort the incidents by department and also separate the lost-time injuries from medical treatment cases. Then use the worksheet shown in Figure 9.1 to assist in identifying the behaviors that are the common thread in the injuries experienced in your facility. This worksheet will also help you identify the practices that occur most frequently. Use the extra space in each category for other behaviors that you find may contribute to injuries.

You should also do an analysis of when employees are most likely to experience an injury, in terms of both the time of day and the day of the week. This information will be critical in designing your process, as explained below. You may also want to collect data on the extent to which behavior contributes to injuries so that you can create a pie chart such that shown in Figure 1.6. Later, in your kickoff meeting or during observer training, you can include this diagram to help explain why you are implementing this new approach to safety.

9.3 DEVELOP A LIST OF CRITICAL SAFE PRACTICES

The next step is to prepare a list of safety practices that will be needed to create your checklists using the information you gathered in the previous step to sort the categories and behaviors in the order of their contribution to injuries. Organize the remaining categories in order of their contribution to injuries, with the category that contributes most at the top of your list. In other words, place the worksheet category that makes the largest contribution to injuries (usually "body position and ergonomics") at the top of your list. Then sort the behaviors within each category to reflect this same order of contribution to past injuries so that you have the behaviors that have historically contributed to the most injuries at the top of each category and those that have contributed the least at the bottom of each category.

Your next step is to compare the lists developed for each department to determine whether they share a common set of critical practices or each department is unique enough to warrant a separate checklist.

VBSP Incident Analysis Worksheet

Department/Area: _____

1. Body Position and Ergonomics	No. of Incidents	Examples/Tasks (Use reference no. on additional sheets)
1.1 Proper body mechanics when lifting or carrying		
1.2 Proper body mechanics when reaching or pulling		
1.3 Clear of "line of fire"		
1.4 Eyes on path		
1.5 Eyes on work		
1.6 Appropriate pace		
1.7 Clear of pinch points		
1.8 Clear of sharp edges		
1.9 Clear of hot surfaces or materials		
1.10 Maintains three points of contact (when ascending or descending stairs, steps, or fixed ladders)		
1.11 Ergonomics/repetitive motion		
1.12 Stays on paths and walkways		
1.13		

2. Tools and Equipment

2.1 Use of scaffolding and portable ladders		
2.2 Selection of tools and equipment		
2.3 Proper use of tools and equipment		
2.4 Condition of tools and equipment		
2.5 Location or storage of tools and equipment		
2.6 Use of vehicles and mobile equipment		
2.7 Guards in place		

Figure 9.1. Worksheet for analyzing past injuries and incidents (continued on next page).

In general, you should start with the assumption that you will need different checklists for different areas. In other words, you should assume that production, maintenance, the laboratory, and the warehouse would each need a separate checklist. In some cases, you may be able to justify having a single checklist. Often, the ideal will be to have a checklist that is roughly 75 percent generic, with a separate section that is specific to each department. Figures 9.2 and 9.3 show examples of the front and back of a generic checklist that might be appropriate for a

DEVELOP A LIST OF CRITICAL SAFE PRACTICES   73

VBSP Incident Analysis Worksheet
(Continued)

3. Work Conditions/Housekeeping	No. of Incidents	Examples/Tasks (Use additional sheets if necessary)
3.1 Use of tags and locks		
3.2 Equipment de-energized		
3.3 Proper permits		
3.4 Work areas free of slip and trip hazards		
3.5 Electrical cords and hoses rolled and stored properly		
3.6 Signs		
3.7 Barricading of hazardous conditions		
3.8 Fire and emergency equipment		
3.9 Adequate lighting		
3.10 Chemicals and materials stored and labeled properly		
3.11		
3.12		

4. Personal Protective Equipment		
4.1 Proper use of respirator/breath apparatus		
4.2 Fall protection		
4.3 Eye/face protection		
4.4 Proper work gloves or hand or arm protection		
4.5 Hard hat		
4.6 Hearing protection		
4.7 Protective clothing		
4.8		
4.9		

Figure 9.1   (*Continued*)

location, while Figure 9.4 shows a checklist with generic and department-specific items.

While the design team is conducting this analysis, you may want to conduct several brainstorming sessions with employees to identify items they consider important to include on the checklist. (Before you begin, you may wish to look ahead to the section on pinpointing in Chapter 21.) When brainstorming, record

Safety Observation Checklist

Observer(s): _____ Date: _____ Time: _____

Team: _____ Location: _____

Instructions: For each safety practice you observe, record a short vertical line "|" in the appropriate column to indicate each safe practice and to indicate each practice that causes you to be concerned about the potential for injury. Keep a tally by recording these in the appropriate space. Describe each significant safety practice and concern in the "Comments" section. Record "Job-specific safe practices" in the appropriate space.

1. Body Position	No. of Safe Practices	No. of Concerns	Comments (Do not use names)
1.1 Body mechanics when lifting, reaching, or pulling			
1.2 Eyes on path or work			
1.3 Clear of pinch points, sharp edges, and hot surfaces			
1.4 Clear of "line of fire"			

2. Work Conditions

2.1 Proper permits			
2.2 Equipment locked out and de-energized			
2.3 Work areas clean and free of slip or trip hazards			
2.4 Storage of materials			
2.5 Signs and barricading			

3. Tools and Equipment

3.1 Use of tools and equipment			
3.2 Use of vehicles and mobile equipment			
3.3 Location of tools			
3.4 Condition of tools and equipment			

4. Personal Protective Equipment

4.1 Fall protection			
4.2 Respiratory protection			
4.3 Job-specific eye and face protection			
4.4 Job-specific hand and arm protection			
4.5 Hearing protection			
4.6 Protective clothing			

Job-Specific Safety Practices

Total			

Figure 9.2. Recommended format for a generic safe-behavior observation checklist.

both behaviors and conditions. Do not be concerned about labels or abstractions at this stage; simply record the employees' ideas. After completing the list, go back over each suggestion to ensure that it identifies a pinpointed behavior or condition. Often this exercise will fit well within an existing safety meeting. You may also want to review your initial list of safe practices with other appropriate employees and managers to get their input and suggestions on what practices or conditions should be added to the list.

Definitions

Body Position

Body Mechanics When Lifting, Reaching, or Pushing	Position body close to load, maintain natural curve in lower back, use mechanical lifting devices when possible, get help when not sure of load or with heavy or bulky load. Square up to task (do not twist at waist). Also consider risk of repetitive-motion injury and discuss "body breaks" when appropriate.
Eyes on Path or Work	Face and head generally pointed in the direction of travel (also looks when stepping down or backwards). Observe path prior to tracing lines. Watching hands engaged in tasks.
Clear of Pinch Points and Sharp Edges	Body parts kept from places where they might be mashed, pinched, cut, or burned (as when freeing jammed equipment or frozen bolts); also fingers kept clear of doors, lids, and covers.
Clear of "Line of Fire"	Not working or standing in the path of equipment or materials that might shift, relieve pressure, move, or fall (e.g., never working in front of a pressure-relief valve).

Work Conditions

Proper Permits	Compliance with all steps required on p.14 of the *Safety Rules and Regulations Handbook*; also welding/burning permits, tagging and flagging, confined space entry, lock out/tag out, blinding, excavations).
Equipment Locked Out and De-energized	Equipment locked out when possible and de-energized of all type of power (electrical, hydraulic, pressure, and mechanical).
Work Areas Clean and Free of Clutter	Walkways and buildings kept clean and free of spills and clutter that might contribute to slips or falls (e.g., hoses and cords routed to avoid crossing walkways when possible; also, coiled and stored when not in use).
Storage of Materials	Chemicals (paints, solvents, lubricants) labeled properly and stored in proper containers and location. Parts and supplies stored properly (e.g., spray cans capped).
Signs and Barricading	Slippery surfaces and hazardous work areas surrounded by barricades, tape, or signs. Comply with signs and barricades.

Tools and Equipment

Use of Tools and Equipment	The right tool used in the manner intended. Tools not substituted for others (e.g., a file not for a pry bar or a screwdriver for a chisel). Guards in place. Tools/equipment deenergized for maintenance (as when changing a saw blade or drill bit). Pneumatic and electrical tools, impacts, grinders, drill masts, jack stands, chain falls, hoists, chokers, overhead cranes, and other lifting devices used only for applicable tasks. Ladder used appropriately: i.e., properly slopes; not standing on top step; tied off when necessary; etc. (See Mobile Equipment, below, for stationary welding and cutting equipment.)
Use of Vehicles and Mobile Equipment	The right equipment used for tasks. Go-devils, forklifts, wench trucks, man-lifts, and other mobile equipment used in a safe manner (e.g., no loads carried on booms, terminals guarded on welding machines, proper caps on cutting rigs, gas turned off for unattended cutting rigs, fire extinguishers available, etc.).
Location of tools and equipment	Tools positioned so as not to create a hazard and stored properly when not in use.
Condition of Tools and Equipment	Tools and equipment maintained in good working order. Guards in place. (Electrical cords free of fraying, wooden handles free of cracks, hammer heads in good shape, welding hoses or leads well-maintained, etc.)

Personal Protective Equipment

Fall Protection	Proper use of handrails, scaffolds, harnesses, life lines, and retrieval systems, and three points of contact maintained at all times.
All Other Items in This Category	Selection, condition, and use of personal protective equipment as described in *Safety Rules and Regulations Handbook*, pp. 18–31. Additional protection used for specific tasks when required such as goggles when working with corrosive chemicals or grinders or a faceplate when welding. Appropriate gloves used when handling solvents and other chemicals, welding, and electrical work.

Job-Specific Practices

Use this section to document job-specific practices or procedures that are not included above.

Figure 9.3. Samples of the operational definitions that would be on the back of the previous checklist.

Finally, go back over the list and change each item to describe the safe behavior rather than the unsafe behavior. Again, you may wish to refer to the section on pinpointing in Chapter 21. You must define the desired safety practices in precise detail so that different observers can agree on how to score a given situation.

Bindery Coaching Checklist

Observer: _____ Day: _____ Date: _____ Time: _____

Instructions: Record the number of times you observe each safety practice and the number of times you had a concern. Check the important positive practices you plan to recognize and significant concerns you plan to discuss, if any. **Do not** record names of persons.

Plant-Wide Pinpoints

Safety Practice	No. Safe	No. of Concerns	Comments (No Names!)
Manual Lifting (41) — Bends knees, keeps back straight, holds load or crane close to body, pivots feet rather than twisting body when manually lifting. Grasps load securely before moving. Only lifts balanced and steady loads. Asks for help or uses lifting equipment for loads > 50 lb.			
Eyes on Path (32) — Focuses eyes in direction of travel to avoid bumping into, tripping over, or slipping on objects or liquids, e.g., when walking, using air crane, or carrying slip sheet. Has a clear line of vision to floor when carrying or pushing an object. Looks at steps, platforms, etc., before stepping up to ensure secure footing. Ties shoelaces. Walks instead of runs.			
Eyes on Task (29) — Focuses eyes on task in which hands are involved. Avoids pinch points, i.e., keeps hands out of equipment during runs, while changing rollers or signatures. Keeps hands out of the line of fire.			
Housekeeping (20) — Picks up and stores or disposes of any unnecessary item on floor and wipes up spilled liquids as soon as observed. Uses absorbent matting. Dries decks of spray mist. Wipes up drips from carrying racks. Keeps walkways free of obstructions. Puts tools and materials away when finished.			
Equipment De-energized (11) — Turns off equipment before performing maintenance or repairs. Activates stop safe before clearing jams. Locks and tags out power before working on equipment.			

Bindery Pinpoints

Safety Practice	No. Safe	No. of Concerns	Comments (No Names!)
Safe Cutting — Cuts away from body and keeps other hand out of blade direction. Has a firm grip on tools (e.g., wrenches) before applying force.			
Right Tool — Uses tools designated for task.			
Guards in Place — Puts guards in place before operating equipment (e.g., protective covers on trimmer knives).			
Eye Protection — Wears glasses with side shields (e.g., when working on trimmer). Wears face shield when grinding.			

8/29/02

Figure 9.4. Example of format for a department-specific safety observation checklist. (Adapted from Quebecor World Hazelton. Used with permission.)

On your checklist you want to include safe practices, not unsafe practices. The rationale is twofold. First, you must be very clear about desired safety practices in order to facilitate communication to everyone in the area. Second, you want to increase the attention being paid to safe behavior and what employees are doing right. A large part of the acceptance of this safety program comes from taking a positive approach versus the more traditional but negative approach of looking for unsafe practices.

As you identify safe behaviors for your checklist, you should also try to eliminate redundant items and items that can be logically combined. For example, your initial list of pinpointed safe practices might include such behaviors as "wearing hard hats," "wearing safety glasses in designated areas," and "wearing hearing protection in high-noise areas." Your design team might combine these behaviors into a broader practice such as "wearing appropriate protective equipment." Or, you may sometimes want to leave such items separate to emphasize the importance of each to the safety of your workplace.

You generally want the items to be mutually exclusive so that your observers are less likely to make errors when recording their observations on checklists. For instance, you may not want checklist items to include both "wearing appropriate protective equipment" and "wearing safety glasses in designated areas." This would create confusion for observers and would make observation data less reliable.

Rule of Thumb: First develop a list of safety practices for each department or area; then see how they can be combined.

9.4 DRAFT AND REVISE CHECKLISTS

Once you have developed the list of pinpointed safety practices, begin the process of designing the actual checklists. The first step is probably to shorten your list of pinpointed items. To do so, consider the following criteria:

- Importance to safety (based on incident records and design team members' opinions)
- Frequency of occurrence
- Observability
- Overlap with other items on the list

Do not eliminate any safety practices that you identified on the basis of serious injuries. If possible, every unsafe act that has caused an incident during the last several years should be addressed by a safe work practice reflected on your final list, although it may be combined with other behaviors under a common safety practice. You should also consider eliminating pinpointed checklist items that (1) have a relatively low risk of minor injury *and* (2) would not have a major impact on performance.

Rule of Thumb: Keep your observation checklists to no more than one side of one sheet of paper.

Note that this discussion has been about developing multiple checklists, not a single checklist. You want the observation data to reflect the responsibilities of a single group or area. Each checklist will be more meaningful for discussion in safety meetings if those participating in the meeting are clearly responsible for the data on it. This usually requires that each checklist needs to reflect a single functional unit from the organizational structure.

Depending on the size and complexity of your organization, you may need several different checklists. Safety practices are different in the laboratory, in the warehouse, on the plant floor, and for maintenance. At the same time, keep the number of checklists as small as practicable. If your organization is not large and complex, you may be able to get by with one or two checklists.

You will need to designate what data need to be collected in the checklist headings. You will also typically want to record the name of the employee conducting the observation, the employee's department, where the observation is being conducted, and the date and time. Other information to be collected as designated by the headings at the top of your form might include the shift, the crew, whether the worker observed was an employee or a contractor, and the number of people observed (if more than one). Be sure to include a brief set of instructions for completing the form. You may also want to allow space, typically at the bottom of the checklist, for additional comments.

Generally, you should not identify the specific work task that is being observed in the header information as that is better included in the observer's comments. If you use a software database to analyze your observation data, you will be able to identify what was being done when a specific concern occurred if observers record the task as part of their comments. Including the task as part of the heading makes this type of analysis more difficult when using most commercially available databases.

Figures 9.3 through 9.6 show several alternative approaches to checklist design. At this point you have to find the right compromise between ease of use, completeness, and level of accuracy.

The checklist seen in Figure 9.2 shows several key features typical of a checklist design that are important to consider in your checklist design. Under "Comments" notice the statement "Do not use names!" to remind observers not to include the names of the employees they observe. Also, in the column headings, notice the wording "No. of Concerns" rather than "No. of Unsafe Practices." The rationale for this wording is to remove the stigma attached to the word *unsafe*, which implies that an employee is breaking a safety rule or doing something wrong. Most employees are uncomfortable about telling their peers they are doing something wrong, but they are more comfortable expressing personal concern for their co-worker's safety and well-being. This language is also consistent with the values identified by the design team and helps demonstrate the alignment of the process with those values in that number of concerns is consistent with a stated value of "concern for fellow employees." It further supports the terms that should be used in discussing the observations (feedback discussions are covered later in this chapter). Other organizations use terms like *At risk* to communicate that an observer considers a co-worker to be engaging in a practice that places him or her at risk of injury.

Also note the length of the checklist in Figure 9.2. This is probably close to the upper limit on the number of safe practices to include on your checklist. In fact, you

Safety Observation Checklist

Observer: _____ Date: _____ Time: _____
Observer's Dept: _____ Area Observed: _____

Instructions: For each safety practice you observe, record a short vertical line "|" in the appropriate column to indicate each safe practice and to indicate each practice that causes you to be concerned about the potential for injury. Keep a tally in the appropriate space. Describe each significant safe practice and concern in the "Comments" section.

1. Body Position	No. Safe	No. of Concerns	3. Tools and Equipment	No. Safe	No. of Concerns
1.1 Body mechanics when lifting, reaching, or pulling			3.1 Use of tools and equipment		
1.2 Eyes on path or work			3.2 Use of vehicles and mobile equipment		
1.3 Clear of pinch points, sharp edges, and hot surfaces			3.3 Location of tools		
1.4 Clear of "line of fire"			3.4 Condition of tools and equipment		
2. Work Conditions			**4. Personal Protective Equipment**		
2.1 Proper permits			4.1 Fall protection		
2.2 Equipment locked out and de-energized			4.2 Respiratory protection		
2.3 Work areas clean and free of slip/trip hazards			4.3 Job-specific eye and face protection		
2.4 Storage of materials			4.4 Job-specific hand and arm protection		
2.5 Signs and barricading			4.5 Hearing protection		

	Behavior*	Comments
Safe Practice	—·—	Description: _____ Potential impact: _____ Contributing factor(s): _____
Concern	—·—	Description: _____ Potential impact: _____ Contributing factor(s): _____
Concern	—·—	Description: _____ Potential impact: _____ Contributing factor(s): _____

*Fill in code corresponding to behavior in top portion, or write "Other"

Figure 9.5. Alternative format for a generic safe-behavior observation checklist.

should probably try to limit the number of critical practices to 12 to 15 or less if possible. As an alternative to having several checklists, you may choose to have a section in which observers add pinpointed safety practices appropriate to the tasks they plan to observe, as shown in the final section of Figure 9.2, entitled "Job-Specific Safety Practices." This section is open ended and observers simply fill in specific safety practices applicable to the tasks they observe or plan to observe.

Safety Observation Checklist

Observer(s): _____ Date: _____ Time: _____

Task(s): _____

Instructions: For each specific safety practice below, record the number of points possible if all employees you observe are 100% safe on that practice. Then record the actual number of points earned on the basis of your observations. Make comments on safe practices as well as any practices that create a risk of injury.

Excavations and Trenching	Wt.	Possible Points	Points Earned	Comments (Do not use names)
Soil placed at least 2 ft from edge	20			
Properly sloped or shored (5 ft or deeper)	10			
Ladder exit within 25 ft (4 ft or deeper)	10			
Vehicles and equipment at least 5 ft from edge	10			

Aerial Basket Operations

Maintain safe distance from AC power	20			
Using voltage tester prior to tasks	20			
Both feet on bottom of basket	5			
Vehicle's rear tires chocked	5			

General Safety

Rescue or fall protection used	10			
Correct body position	10			
Proper use of tools	5			
Proper use of heavy equipment	5			
Worksite protected with signs, cones, barricades, etc.	5			

Personal Protective Equipment

Hard hats	5			
Job-specific eye/face protection	5			
Job-specific gloves	5			
Hearing protection	5			
Company-approved footwear	5			
Total				

$$\frac{\_\_\_ \text{No. Points Earned}}{\_\_\_ \text{No. Possible Points}} \times 100 = \_\_\_ \% \text{ Safe}$$

Figure 9.6. Alternative format for a safe-behavior observation checklist using a point system.

Observers may refer to written job procedures, task hazard evaluations, material safety data sheets, or other employees to help define the specific safety practices appropriate for the tasks being performed during their scheduled observations.

Figure 9.3 shows the back of the checklist in Figure 9.2. You need to provide the observers with clear operational definitions of the safety practices that you are targeting on your checklist. Your definitions should include a pinpointed description of the practices on the front of your checklist (see the section on pinpointing in Chapter 21). The pinpointed definitions should include examples, and whenever possible the examples should come from incident investigations. Such definitions will ensure that the behaviors that have caused problems in the past are clearly addressed in the observation process. You can later add additional examples should other incidents or near misses occur. You may want to place your examples in parentheses throughout this page. Including actual examples is another way in which observers come to see the checklists as relevant to your organization.

Including the definitions on the back of the checklist additionally ensures that they are available to observers for easy reference. Some organizations laminate the definitions onto clipboards that the observers use when conducting observations.

Having a multipage list of definitions separate from the checklist is not a good idea, as they will not be available for easy reference. Some organizations have kept them separate with the rationale that it forces observers to learn the definitions. However, observers learn more from conducting the observation, and having the definitions readily available for easy reference contributes to more reliable data collection. Further, some observers will simply make their best guess about where to score a particular practice, seldom taking the time to look elsewhere for the applicable definition.

Figure 9.4 shows an example of a department-specific checklist that combines a sitewide section with a department-specific section. Notice that this organization elected not to use the standard categories given in the incident analysis worksheet seen in Figure 9.1. Its checklist is organized simply into two categories: sitewide pinpoints and bindery pinpoints (which are the items specific to the bindery department).

The checklist in Figure 9.4 also illustrates another strategy worth considering in the construction of your checklists. It includes the number of injuries that could have been prevented through each safety practice, which helps employees to realize that these items are important to safety. Generally, including the actual number of injuries is more effective than using percentages because the latter are often fairly small when spread across all of the practices on your checklists; thus percentages seem to trivialize the contribution of each practice in often making some of them seem insignificant.

The checklist in Figure 9.5 illustrates yet another option in how you may choose to format your checklist. The advantage of this format is that it more clearly communicates the expectation that observers will find three practices, one positive and one or two of concern, to comment on during their observations.

Figure 9.6 shows a checklist that uses a point system to weight the importance of the different safety practices. This checklist is designed in such a way that all the practices will usually not be relevant during a single observation and so only what is observed will be converted into a percentage using the equation at the bottom of the figure. If all of the practices on the checklist will be observed each time observers conduct an observation, then a simpler approach would be to create the weighting by allocating 100 points across the practices on the checklist. Observers can then simply add the points at the end of their observation to calculate an overall percent safe. This approach is not appropriate for most organizations that do not require an overall percent safe for each individual observation.

TABLE 9.2. Questions That Need to Be Answered When Constructing Observation Checklist

Question	Yes	No
1. Is the checklist no longer than one page?		
2. Are all items clear and specific?		
3. Are all items mutually exclusive of other items?		
4. Are the instructions clear and concise?		
5. Is a space provided for the observer's name?		
6. Is a space provided to record the date and time of observations?		
7. Are operation definitions available for each item?		
8. Do your definitions include examples from your incident investigations?		

You may choose to count the occurrence of safe practices and concerns while scoring the physical condition of different work areas as safe or unsafe for other items. An empirical study by Sulzer-Azaroff et al. (1990) used an observation sheet that combined these approaches. On the upper half of their observation sheet, they counted the frequency of safe behaviors related to compliance with procedure and the use of safety equipment. On the lower half of the sheet, they scored the condition of various zones of their facility as either safe or unsafe based on the presence or absence of several pinpointed safety hazards. They included a simple layout so that observers could mark the location of hazards along with space to describe the hazard.

Do not be concerned about designing a perfect checklist. Developing a good checklist is an empirical process and you always learn more about them as your experience increases. Your best strategy is to experiment for several weeks, then design your best observation checklist and use it for three to six months. Remember that the checklist is a dynamic tool. It should change to meet the changing needs of your work environment, yet should remain stable for periods long enough to document goal achievement, generally from six months to a year.

Table 9.2 summarizes considerations for constructing your observation checklists.

9.5 DEVELOP THE OBSERVATION PROCEDURE

Your design team should develop formal guidelines for conducting observations. You may wish to consider flowcharting the procedure to ensure that you have a logical sequence of events everyone can understand. Here are some questions your design team should consider:

- Who will conduct the observations?
- Is being an observer voluntary?

- Is being observed voluntary?
- Should the observer announce the observation?
- How often will observers conduct observations?
- When will observations be conducted?
- Will observations occur across or only within departments?
- Will employees observe an area, a single employee, or specific tasks?
- How will you handle contract personnel?
- Where do observers get observation checklists and how do they submit completed observations?

The remainder of this section will discuss some design options and considerations for answering each of these questions. After your design team has resolved them, you should draft a written procedure with step-by-step instructions for each of these components.

Who will conduct the observations? In developing the observation procedure, the first question to address is who will conduct the observations. Currently, almost all organizations plan to involve employees in conducting observations, and the data clearly support involving all employees in conducting observations. Ultimately, most organizations should strive to involve all employees in conducting safety observations. Still, for some organizations, getting managers and supervisors involved in conducting safety observations, at least as an initial step, is often an appropriate start. Figure 9.7 summarizes considerations for deciding who should conduct observations.

While having a dedicated group of observers may seem a good idea and may even be appropriate for some organizations, it often creates an "us-versus-them" divisiveness that does not promote the sense that safety is everyone's responsibility. In the worst case, other employees consider the observers to be a sort of "safety police," and the effect is almost the same as adding an additional layer of supervision within the organization.

If employees are to conduct the observations, steering committee members may initially conduct observations with employees who have volunteered to participate in conducting them as soon as they are trained in doing so. Remember, participation in the design team is based on who it is determined should take this responsibility. If observations are to be a management responsibility, your design team members should primarily be managers and supervisors. If observations are to be an employee responsibility, the design team should involve representative employees.

Is being an observer voluntary? For most organizations that implement behavioral safety, participation as an observer is voluntary for employees, but this is a decision that should be made by the design team. In some organizations, conducting observations is voluntary for employees but is required of managers, supervisors, and steering committee members.

Some organizations implement programs in which conducting observations is a job requirement. The problem with this approach is that it can increase the number of forms that are completed as a paperwork exercise without actual observations.

STEP 2: CREATING THE SAFETY OBSERVATION PROCESS

If your organization is:	Then:	Considerations
Typical of most organizations	All employees should conduct observations	Ensures a high level of understanding of safety requirements Requires management effort to maintain observations Usually required for daily observations
Still working on the design and implementation of a behavioral safety process	Design team members should conduct observations	Ensures a realistic process Provides a base for training others
A traditional chain of command with management that has not routinely shown a commitment to safety	Managers and supervisors should conduct routine safety observations	Supervisors develop a good understanding of the process Employees will not get the benefit that comes from being observers Often easier to maintain Works well for weekly or monthly observations Often important in building readiness
A traditional chain of command with evolving involvement in safety or quality team process	Start with management, supervision, and representative employees, then begin involving all employees	Important to have both employees and managers participate on the design team
A traditional chain of command with evolving involvement in safety or quality team process, but low trust	Employees should conduct observations without involvement of management in the behavioral process	Managers and supervisors should continue to perform routine audits and observations but outside of the formal behavioral process Not ideal as it tends to support an "us vs. them" mentality Typically should be a step toward full involvement

Figure 9.7 Considerations for deciding who should conduct observations.

The steering committee then has to address this problem by developing plans to encourage quality observations, and tracking the quality of observations is difficult. On the other hand, if conducting observations is voluntary, the steering committee has only to develop action plans that encourage employees to conduct them. While this is a challenge, measuring participation is much easier than measuring the quality of the observations.

If the process is voluntary, the organization has a way of measuring complacency. When employees begin to get complacent about safety, they stop conducting observations. These organizations must then take action to promote safety awareness and get employees involved in safety improvement efforts. When observations are mandatory, the organization will often continue to get observation forms and be unaware of a growing complacency toward safety.

For these reasons, a voluntary process is most appropriate for the employees of most organizations. However, because participation by managers, supervisors, and steering committee members is highly important, most organizations should make their participation a requirement.

Is being observed voluntary? Being observed generally should be considered a job expectation in most organizations that implement behavioral safety. In the early days of behavior-based safety, observers in some organizations asked permission of co-workers before conducting the observation. The problem with this approach is that it effectively allowed employees to opt out of participation in the company's safety improvement efforts. When this question gets asked formally, employees that otherwise would have participated elect not to allow the observation. Still, this is an option that the design team should consider, and it may be appropriate for organizations with very low levels of trust between employees and management. Often, in organizations with low trust, a better option is to make being observed a job expectation but to allow employees to opt out of a particular observation. Thus an employee has the right to refuse a particular observation but not to refuse to allow observations altogether. This gives employees the freedom to choose when observations occur and allows them to opt out if they have a personality conflict with a particular observer, if they are in a hurry and do not want to take the time to discuss the observation, or perhaps if they think an observation at another time would have more value. Providing this option also often helps overcome resistance to the observation process and is therefore appropriate for some organizations.

Should the observer announce the observation? You also need to consider whether observers should announce when observations are going to be conducted. For most organizations, announcing safety observations is an important practice in support of the values of openness and respect for employees. An observer can easily announce the observation when he or she enters the work area, often by simply making eye contact with the coworker to be observed, then holding up a clipboard and pointing at it prior to starting the observation. Or the observer can schedule the observations of co-workers during a Monday toolbox safety meeting, for instance, or at the start of work. The observer might simply mention something like "I'll be coming out to visit your job site sometime later this morning to conduct a safety observation" or "We'll probably do an observation of the maintenance shops on Thursday afternoon."

Obviously, the disadvantage of this approach is that observers may not observe a work sample typical of normal practices. Social psychology studies have revealed that people act differently when they know they are under observation. This is not really as large a problem as it may seem. If employees work more safely during observations, observers have the opportunity to reinforce safety practices that are more likely to be performed in the future. Also, when the observation is announced, concerns that are noted will be more important issues for discussion because the employee is not aware of these issues. Thus such concerns are more likely to be true training problems when employees do not know what they should be doing to protect themselves from injury. Some organizations may, however, choose unannounced observations in order to ensure truly representative data on compliance with safety procedures. You may want to consider announcing observations initially, and then, when employees understand and trust the process, begin unannounced observations.

How often will observers conduct observations? The frequency of observations is important. The risk associated with your business should determine whether observations are made daily, weekly, or monthly. If you are in a high-risk business with many employees, you should probably conduct daily observations. Most manufacturing organizations will want to conduct weekly observations. You may also choose a different frequency of observations for different work areas or levels. You might require supervisors to conduct weekly observations of their work areas, for example, while upper level managers and staff members conduct monthly observations. Figure 9.8 summarizes these considerations.

When will observations be conducted? The observers themselves should generally decide when to conduct observations. Accordingly, you might ask supervisors to conduct their observations during the week without specifying when they

If employees in your organization:	Consider conducting observations:	Considerations
Are at moderate to high risk of an injury incident	Daily or weekly	Frequent observations increase the rationale for involving employees in observations of their work areas May decrease frequency as an area improves its safety record
Are at low to moderate risk of an injury incident	Weekly or monthly	This is a good frequency for observations conducted by managers and supervisors Most empirical studies use weekly observations
Have very low risk of an injury incident (office workers, for example)	Monthly or periodic, if at all	May need greater frequency to ensure consistency of process Can be useful to develop an understanding of ergonomic issues in an office environment

Figure 9.8. Guidelines for deciding on the appropriate frequency of observations.

should do so. However, it is important to vary the time and day that observations are conducted instead of scheduling them to occur at some fixed time every time. You do not want employees to be able to predict when observations will occur.

You may want to schedule the observations for times when incidents are more likely to occur. Data on when incidents are most likely from your analysis of incidents and injuries should provide guidelines on when observations will have the most value. You may also want to create a separate observation process for special events, such as plant turnarounds or new construction within the work area.

Will observations occur across or only within departments? Employees will usually be most comfortable observing jobs they are familiar with, so many organizations start by encouraging observations within observers' departments. Then, as their observation process matures, the steering committee begins to encourage these employees to conduct observations in other areas. In the case of particularly high-risk work areas, observers should find a steering committee member, supervisor, or someone else who is familiar with the work to accompany them when they first enter the unfamiliar area. This approach has a couple of advantages. The employee being observed has someone who can ensure the safety of the observer as well as someone who can discuss issues that the observer has questions about. This approach also makes observers more comfortable when they begin doing observations outside of their own work areas.

Will employees observe an area, a single employee, or specific tasks? The best answer to this question will depend on the nature of your organization and the kind of incidents that your organization has experienced. For manufacturing observations that have many employees working in an area, conducting an area observation will often be most effective wherein observers can observe and provide feedback to several employees in a single observation using a single checklist. For many process plants, on the other hand, observers have to "go where the action is" and thus leave the control room or another vantage point to observe co-workers who are performing work activities somewhere within their unit. Also, the information collected when you analyzed past incidents may have indicated that injuries are particularly likely when employees are performing specific tasks, in which case the observation process needs to ensure that employees observe those tasks. Your analysis should also have told you whether incidents are more likely during routine operations or during upset conditions. Obviously, you need to plan your observations accordingly because you want employees to do observations when incidents are most likely to occur.

Sometimes you will want to address this question based on the nature of the activities identified on the observation checklist. You will often have a combination of different types of observations on the same observation sheet. If "forklift boom in proper position" is on a warehouse observation form, then you would want to try to stay in an area until you observed forklift operations. Depending on the level of activity, the observer might wait a few minutes to observe this task or simply score this item as "not applicable." You might establish a simple guideline, such as "Try to observe the occurrence of a checklist activity (or a completed task sequence) if you can do so in less than five minutes."

88   STEP 2: CREATING THE SAFETY OBSERVATION PROCESS

How will you handle contract personnel? Many of today's companies make ongoing use of contract personnel. If your organization uses them, your design team should consider whether to involve them. The options for doing so include the following:

- Working with them to establish their own observation process
- Involving them in the design team and making observations
- Observing their activities but not involving them in conducting observations

As a rule, if you have contract personnel onsite on an ongoing basis, you should find a way to involve them at an appropriate level. Probably your best option is to observe them, then provide feedback as you would any other employee. In some cases you may want to be more aggressive with contract personnel than with your own employees. For example, you might be less likely to announce observations in advance, and you may not give these employees the right to opt out of observations. In some cases you may also want to provide the feedback to the contractor employees' supervisor.

Where do observers get observation checklists and how do they submit completed observations? You will also need to plan the administrative aspects of supporting observations, as in where employees obtain blank observation checklists and, after the observations are completed, how they submit the checklists to the steering committee. As part of the data feedback process described in the next chapter, you may want to establish a safety bulletin board in each area. This board can display the blank checklist, and it is a good area in which to locate drop boxes for submitting completed observations. In some cases employees can submit completed observations through interoffice mail, and in others the observers can enter their observation data into a computerized database.

9.6   FEEDBACK ON OBSERVATIONS

As a general rule, you should plan a process in which observers routinely provide immediate feedback as part of their observations. Some practitioners have suggested that the data collection (observation) process and intervention (feedback) process should be separate. The problem with this approach is that encouraging a sufficient feedback is difficult if it is not an integral part of the observation process. Sometimes the two are separated when in a less formal observation process employees carry a data collection card in a shirt pocket for use any time they observe other employees. In addition to a difficulty in prompting feedback, this approach to data collection often results in an undue focus on unsafe acts as it is often just such an act that prompts an employee to record what is observed. Once a focus on the negative begins to characterize the process, however, employees begin to view observations as a "secret police–like" strategy.

For these reasons, observations and feedback should generally be integrated with one another. Observers should provide feedback almost every time they complete an observation. Exceptions to this rule would be any time that stopping to discuss an observation would create a hazard, any time an observation is conducted in a high-noise environment, or during special situations such as when both employee and observer are wearing breathing apparatus that would make such discussions difficult. In these cases, the feedback should be given as soon as practical. Other than in situations such as these, observations and feedback should go hand in hand. An observation without feedback is like a gift given and received without a "Thank you."

How should observers provide feedback on their observations? The discussion of the observation should be a dialogue between the observer and the employee(s) observed. This feedback procedure should be an important element of the training provided to all observers. The observer should summarize the significant safety practices that were observed, then communicate the one or two practices that caused the greatest concern. Generally, the observer begins by listing one or two of the things that the employee was doing right: "Here are some ways that you are minimizing your risk of injury... (lists the relevant safe practices)." Or the observer might use a variation such as "Some practices that I thought contributed to your safety are...." Our observer next goes on to summarize the unsafe practices: "Here are some things I am concerned about... (lists relevant unsafe practices)." Or again a variation such as "These are a few practices that are placing people at risk... (lists unsafe practices)." Or "How would it work if... (followed by a request for the safe practice)."

When giving feedback, a *three-step* feedback process works equally well for both positive and corrective feedback:

1. Describe the behavior that was observed.
2. Discuss the potential impact on the employee and co-workers.
3. *Listen to what the employee has to say* and then, for corrective feedback, make a specific suggestion that pinpoints what he or she should be doing differently.

Using this model to discuss a safe practice, the observer might say something like "I noticed you used proper lifting techniques in lifting with your legs while keeping your back straight. Lifting like that will help you keep your back healthy." Then, pause and listen to any response from the employee. Similarly, when discussing an area of concern, the observer might say, "I was concerned about the fact that you were not wearing a harness or any other form of fall protection. At six feet off the floor, you're high enough to suffer a serious injury if you were to fall." Then listen to the employee's response.

An employee who has a reason for an at-risk behavior is likely to tell the observer at this point. In the above example, the employee might respond, "Yeah, I stopped by the storeroom, but they didn't have a harness available and I needed to get this job done." When employees have such a reason for their action, the observer should record it on the observation form as it will be important in helping the steering committee in developing action plans to address that concern.

While getting such information is important, observers should be taught not to interrogate the employees that they observe. The discussion should be a two-way, problem-solving discussion that is educational for both parties. In particular, observers should avoid two kinds of problems caused by (1) asking rhetorical questions and (2) asking questions that start with "why." The problem with rhetorical questions is very simply that they tend to make employees angry. This is the problem with the *two-question method* often taught to observers in Du Pont's Safety Training Observation Process (STOP). In this procedure observers ask two questions. First they ask, what could happen. This prompts the employee to identify the risk in the situation and thus respond to the same aspects of the situation that the observer is inquiring about. Next they ask how the risk could be avoided, which prompts the employee to identify the appropriate safe practice or corrective action. Recent research on leadership suggests that such questions tend to generate anger rather than the hoped-for educational impact.

Asking "Why...(is an employee doing something)?" also has a negative impact in that it tends to make the employee defensive rather than set the occasion for a constructive problem-solving discussion. The same information can always be gathered by asking questions that start with "what" or "how," which does not seem to create the same defensiveness. Asking a question such as "What are the barriers that prevent you from using a harness?" creates much less defensiveness than "Why aren't you wearing fall protection?"

If your observers will be conducting observations of small groups of employees, they might provide feedback to the group using basically the same technique. If the group is working on a common task, the observer can usually give feedback to the group. If group members are working on different tasks but in the same general area, the observer should usually give individual feedback. Once the observer has completed the observation, the observer might approach the group, review the things that the group was doing well, then discuss problem areas by saying, "Here are some things I have concerns about..." before reviewing the unsafe practices he or she observed. Notice that the observation checklist in Figure 9.2 has columns for number of safe practices and number of concerns followed by a column for comments. The form then prompts the kind of discussion planned by the design team consistent with one of the stated values, "concern for fellow employees."

In addition, you should provide guidelines on how to handle disagreement about a safety practice. If a disagreement arises, the observer should always try to reach a consensus with the employees who were observed. If that fails, the question should be referred to the steering committee for resolution.

Some Common Configurations. Organizations need to design their behavioral safety processes to meet their unique needs. Many issues, including factors such as the size and complexity and the history of involvement of employees in safety improvement efforts, affect the final design. Figure 9.9 presents some common configurations for behavioral safety processes.

	Ideal	Employee Based	Empirical Studies	Management Based	Staff Based
Observers	All employees (including supervisors and managers)	Employees only (either all employees or a dedicated group of observers)	Independent (often from outside organization)	Managers and/or supervisors	Staff personnel (usually safety staff but senior maintenance person may observe conditions)
Frequency	Weekly	Daily or weekly	Weekly	Weekly	Weekly or monthly
Feedback	Behavioral feedback during observations Data reviewed in safety meetings	Behavioral feedback during observations Data reviewed in safety meetings	Data summaries to manager Data reviewed in safety meetings	Behavioral feedback during observations Data reviewed in safety meetings	Data summaries to manager Data reviewed in safety meetings
Observations Announced?	Usually (at least in early stages)	Usually (at least in early stages)	No	For some programs	No
Role of Management	Approve plan Participate in kickoff meetings Review observation data Conduct observations Approve and participate in safety award program	Approve plan Participate in kickoff meetings Review observation data Approve and participate in safety award program	Approve plan	Approve plan Participate in kickoff meetings Review completion of observations Approve and participate in safety award program	Approve plan Participate in kickoff meetings Review completion of observations Approve and participate in safety award program

Figure 9.9. Common configurations of behavioral safety processes.

92 STEP 2: CREATING THE SAFETY OBSERVATION PROCESS

9.7 TRIAL RUN THE OBSERVATION CHECKLIST AND PROCESS

At this point, your design team is ready to do a trial run of your observation process. This trial run has three objectives:

1. Fine tune the checklist and observation procedure to ensure ease of use.
2. Ensure the reliability of the observation procedure.
3. Develop baseline data for each checklist (i.e., for each separate area or function that will observed).

You should walk through the entire observation process and debug each step. Be prepared to go through several revisions of each checklist and procedure.

At this stage, you should also work on developing a reliable observation procedure. Have two observers go through the steps side by side. They should first go through the observation procedure collaborating on how to complete the checklist. After this *calibration training*, at their next observation session they should again score the work areas simultaneously, this time without discussion. Then their checklists should be compared and a reliability coefficient calculated using the following formula:

$$\text{Reliability coefficient (\% agreement)} = \frac{\text{number of items in agreement}}{\text{number of items in agreement} + \text{number of items in disagreement}}$$

For this kind of application, you should strive for a reliability coefficient of 80 percent or better. The better your reliability at this stage, the easier it will later be to train observers. If you have a very clear and simple checklist with well-pinpointed items, new observers will be able to conduct their observations with very little observer training.

Save the data you collect during this phase, especially as you finalize your checklist. It will provide a baseline for evaluating future efforts. It may also provide a basis for setting improvement goals and problem solving in later stages of implementation.

Before the management review, the design team should review its procedure or flowchart, brainstorm what could go wrong at each step, and discuss how to refine the process to minimize the likelihood of each potential problem.

The following table provides clarification on the next steps:

If Your Design Team is Conducting:	Then You May Choose To:
Marathon planning meetings	Skip the section that follows below on management review and go on to step 3, develop feedback and involvement procedures.
Periodic planning meetings	Conduct a review with management as described below, then proceed to step 5, conduct training and kickoff meetings in each area.

9.8 CONDUCT MANAGEMENT REVIEW

Once you have completed your trial run and made final revisions to both your checklist and procedure, the design team should present the observation process to management. This step should be one of the milestones on your initial schedule. Use this meeting as an opportunity for

- management to provide input to the safety process,
- safety team members to be recognized for their participation, and
- the design team to obtain approval to implement the observation process.

All design team members who want to participate should be involved in this presentation. The recognition from management should reinforce their participation in the safety process.

Getting management's input and suggestions on your observation system during this presentation helps ensure it will support your implementation efforts. This meeting also provides an opportunity for ensuring that management understands each component of the behavioral safety process at each stage of its implementation.

Figure 9.10 presents a typical agenda for such a meeting.

- Purpose: Obtain management approval to implement observation system.
- Background on how the design team came into being
- Objectives of the observation system
- Process for developing the observation system
    - List of initial pinpoints
    - Final checklist
- Observation procedure
- Management's role
- Next steps
    - Kickoff meetings with employees
    - Developing feedback and involvement process
- Discussion and management input

Figure 9.10. Typical agenda for presenting observation system to management.

10 Step 3: Designing Feedback and Involvement Procedures

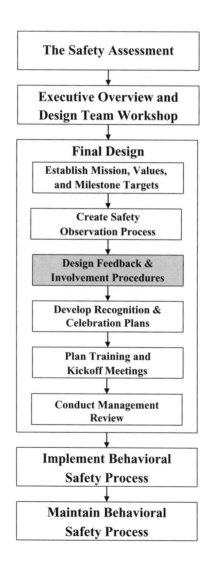

Creating an effective feedback and involvement process includes the following implementation tasks:

Task Number	Activity
1	Develop guidelines for using graphs.
2	Plan reviews of safety process data: • In safety meetings (observation data) • In management meetings (data on the percent of observations completed)
3	Develop guidelines for setting improvement goals.
4	Establish guidelines to expand involvement in observations.

10.1 DEVELOP GUIDELINES FOR USING GRAPHS

Your design team should develop guidelines for using graphs to providing feedback. Numerous studies have shown that posting performance graphs has a significant positive impact on employee performance (Andrasik, 1979), and most research studies investigating behavioral safety programs have included graphs of the observation data. Graphing safety observation data is a good way to communicate the data to employees. In addition, graphs help employees identify trends and set improvement targets. In contrast, some companies have had bad experiences with poorly planned efforts to graph quality data. Such companies may want to avoid the negative associations that some of their employees have with posted graphs.

Figure 10.1 provides guidelines for using graphs.

If:	Then:	And:
Your employees have experience using data to self-manage their work areas and have a good relationship with management	Post graphs in work areas	Review the graphs in weekly safety meetings
Your employees have had a bad past experience with graphs - or - Your employees have a poor relationship with supervision or management	Don't post graphs	Steering committees should review graphs during their meetings and introduce those graphs to employee safety meetings only after employees are comfortable with the observation process

Figure 10.1. Considerations for the use of graphs.

96 STEP 3: DESIGNING FEEDBACK AND INVOLVEMENT PROCEDURES

In developing guidelines for using graphs, you should consider making suggestions on what to graph, provide sample forms for graphs, and provide suggestions on their use. Your design team should consider recommending two graphs for each area, one showing "percent safe" observation data and the other showing the percentage of observations completed each week, as in Figures 10.2 and 10.3, respectively. The graph of percent safe either can be over all observations or it may track a single behavior that the area is targeting for improvement. Sometimes, the steering committee may decide to post a bar graph that shows the percent safe on each practice on the checklist. This graph would be appropriate, for example, when the steering committee wants to share the data with area employees so that everyone knows the basis for the priorities and action plans being developed by the steering committee. The steering committee may either keep the graphs by hand or use computers. The most important requirement is that all graphs be simple and easy to understand.

> **Rule of Thumb for Posted Graph:** The graph should be easy enough to understand that an employee can glance at it when walking by and be able to interpret it without breaking stride.

Note that the graph for recording the observation data also has space for recording when the observation was conducted and by whom. You will want to suggest that each steering committee have a separate graph for observation data from its own area, meaning you may have several graphs on the same scoreboard depending on how you have designed the observation process. For example, you may want each shift to have a separate graph of weekly safety observations that occurred specifically on that shift.

Initially, depending on the sophistication of your workforce, asking observers to update the graphs manually is preferable to generating them by computer. Requiring that observers record the data on the graphs ensures that they understand the data being presented. The feedback will also be available without the delay that often results from a computerized process.

Entering the data into a database or spreadsheet does have some advantages, however. It enables you to easily generate summary reports for distribution, which is particularly important for tracking the percentage of observations conducted in a large organization. Computers can also easily generate monthly or weekly reports on the percentage of observations completed in each area for review in steering committee meetings.

If you recommend that your organization post safety graphs in the work areas, be sure you provide guidelines to all managers and supervisors on how to use such graphs effectively.

> **Rule of Thumb 1 on Using Data:** All managers and supervisors should be instructed *not* to respond in any way to a low percent safe or safety index on graphs. Managers must not put pressure on employees to improve these numbers. They may problem solve the observation data with employees

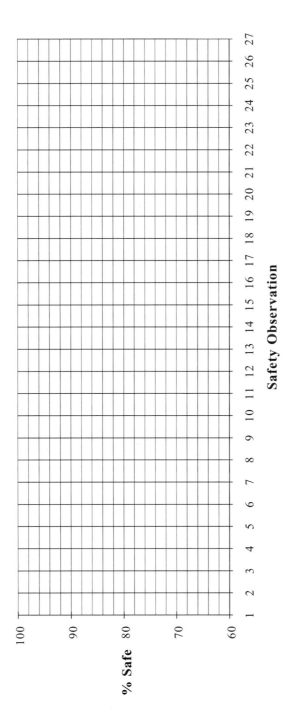

Figure 10.2. Sample graph for recording observation data.

97

98 STEP 3: DESIGNING FEEDBACK AND INVOLVEMENT PROCEDURES

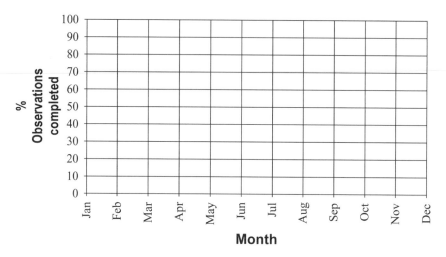

Figure 10.3. Sample graph for data on completed observations.

within the context of appropriate safety meetings but should refrain from any negative or critical comments.

Rule of Thumb 2 on Using Data: All managers and supervisors should be encouraged to make positive comments on good percent safe or safety index scores or improvements evidenced on such graphs. They should also make positive and appreciative comments to appropriate employees both during the workday and during meetings.

In planning to use graphs, you must emphasize safety on the job, not numbers on graphs. The risk in using graphs is placing too much emphasis on the numbers. The result is that you influence reporting, not performance.

10.2 PLAN REVIEWS OF SAFETY PROCESS DATA

To get the maximum benefit from the observation process, you must ensure that the organization makes use of the data. In other words, the data must be reviewed and employees must respond to it. The best way to ensure that people look at the observation data is to build a review of the data into existing meetings. Ideally, both the graphs and observation sheets should be reviewed as one of the first agenda items in weekly safety meetings.

Data on the percentage of observations completed should be reviewed both by the steering committee and in management meetings. Management should focus on managing the safety process, not the results of the process. If management attempts to manage the results of the observation process, it will ruin the integrity of the system. Such pressure from management will eventually bias the observation process and destroy the value of the data.

This is not to suggest that managers and supervisors should forego attention to safety on the job. The observation data are a measure of the extent to which employees work safely. Managers and supervisors must provide daily feedback to employees for on-the-job safety to ensure the success of the behavioral safety process. However, they should emphasize safety on the job and maintaining scheduled observations, not the percent safe resulting from completed observations.

10.3 DEVELOP GUIDELINES FOR SETTING IMPROVEMENT GOALS

The design team should provide guidelines for establishing improvement goals based on the observation data. Ideally, as part of the data review process in safety meetings, employees should establish improvement targets for their area's percent safe. These improvement targets should be based realistically on the existing level of safety practices as indicated by the observations. The goal should be for some fixed time period, such as the next one to three months. The goal is best set for some fairly short period of time, not for the entire year, so that steering committees can make frequent corrections to the process and have regular opportunities to celebrate success. Once the target is agreed upon, each steering committee should draw a goal line on its safety graph with a colored marker so that all employees can tell where they stand relative to the target. Then the employees should identify which safety practices they are going to work on in order to reach their target.

An alternative goal-setting strategy is to establish *process goals* that focus on improving specific safety practices. In other words, rather than setting a specific percent safe, the employees might target achieving 100 percent for the coming week on a specific safety practice or set of practices identified as needing improvement on the basis of observations completed during the previous week.

To ensure that the goals get set, the design team should ensure that responsibility for setting the improvement target is clearly assigned and communicated. In most cases, the steering committee should set safety improvement targets or goals. This procedure would be appropriate if the observation process combined several shifts or work areas that each had separate safety meetings. For other organizations, the person responsible for leading the safety meetings should take responsibility for ensuring that each team of employees sets a safety goal. While employee participation in goal setting is consistent with the philosophy of an employee-driven process, current data do not show a significant advantage for employee participation in goal setting versus goals assigned by management, although employees do like to participate in setting goals (Fellner and Sulzer-Azaroff, 1985). Thus, for other organizations, the design team might plan for management to take responsibility for reviewing the data and setting improvement targets. Figure 10.4 summarizes the options and considerations for goal setting. Obviously, many organizations will have some combination of these.

Establishing this improvement goal is important for several reasons. Progress toward an explicit goal provides a positive source of motivation and helps build pride in the area's safety efforts. In addition, goal setting helps reduce competition by providing a noncompetitive standard of comparison. An effective goal or

Targets set by:	Would typically include:	Considerations:
Natural work groups (groups of employees that typically meet together for safety meetings)	Safety targets: Top 1–2 concerns (improving % safe on that behavior) Process targets: # of observations % participation in conducting observations	Most appropriate when many important issues are behavioral or conditions that can be addressed at the local level
Management team	Safety targets: Top 1–2 concerns (improving % safe on that behavior) Process targets: # of observations by managers & supervisors Participation by managers & supervisors	Most appropriate for addressing facility issues and ensuring visible management support May be the best option when steering committee meeting time is limited
Steering committee	Safety targets: Top 1–2 concerns (improving % safe on that behavior) Process targets: # of observations % participation in conducting observations	Best overall for most organizations—integrates action plans to address target behaviors and facilities related to those behaviors Adds to the time required for steering committee meetings

Figure 10.4. Considerations and options for setting improvement targets.

performance target gives the area team a standard for evaluating its performance. Members can compare their performance with their own goal rather than where they stand relative to other groups. Downplaying competition is particularly important because different work areas may have very different risks and safety requirements. Leading a group of employees toward a common goal is a far better process for building teamwork and cooperation. Chapter 22 outlines the steering committee's responsibilities in this area in much more detail.

10.4 ESTABLISH GUIDELINES TO EXPAND INVOLVEMENT IN OBSERVATIONS

Once the steering committee has the observation process up and running, it will usually want to expand employee involvement. The design team will often set a long-term goal of training all employees as observers, even though the initial design may have placed this responsibility on management and supervision. If you initially established the observation process as a management or steering committee responsibility, suggest to whichever it may pertain that employees should be encouraged to participate three months or so after the new safety process is underway. Expanded participation might begin through joint observations, the new observers participating alongside of supervisors or experienced observers. Just make sure that you provide adequate training and orientation for new observers as described in the previous section (also see the discussion on training new observers in step 5, plan training and kickoff meetings).

10.5 CHECKLIST FOR PLANNING FEEDBACK AND INVOLVEMENT

Figure 10.5 is a checklist to use in planning feedback and involvement.

Does your design team s plan include recommendations on:	Yes	No
1. Posting graphs and copies of the last completed observation data sheet in work areas?		
2. Graphs and observation data to be reviewed in weekly safety meetings?		
3. The review of data on the "percent of observations complete" in area safety meetings and regular management meetings?		
4. A process for setting area improvement targets based on observation data?		
5. Increasing involvement in the observation process?		

Figure 10.5. Checklist for planning feedback and involvement.

11 Step 4: Developing Recognition and Celebration Plans

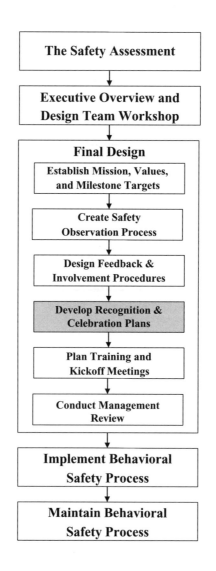

There are three basic options in designing effective safety incentives:

1. Create a safety award process.
2. Support the safety process through an existing compensation process.
3. Provide incentive compensation based on the safety process.

As stated earlier, traditional safety award programs often reward people who take chances or encourage employees not to report incidents accurately. Too many people simply roll the dice. The chance of injury is usually low enough that they do not get hurt even though they take chances. In award programs based on going a fixed time period without an incident, such employees usually get the same award as employees who always comply with safety procedures. Furthermore, if the award is significant, and particularly if the award is significant to a group of employees, such programs tend to discourage the honest reporting of minor incidents by most employees.

To avoid such shortcomings, safety awards and incentives should be based primarily on behaviors that promote safety, such as conducting observations, leading safety meetings, and other activities that directly or indirectly contribute to the safety and well-being of co-workers. In addition, small awards can be provided for maintaining a safe workplace as measured by observation data, perhaps in combination with no lost-workday cases. The safety award process provides a way of celebrating successes and expressing appreciation to employees who work safely as well as those who make special contributions.

> **Safety Awards—Rule of Thumb 1:** Provide safety awards for safe behavior on the job and for activities related to maintaining the safety process (e.g., observations, conducting safety meetings, setting safety goals).
>
> **Safety Awards—Rule of Thumb 2:** Keep safety awards and incentives small. Your awards should be significant enough to support compliance but not significant enough to generate false reporting of safety data.

The second rule is especially critical, as illustrated by the court decision mentioned in Chapter 1 where an employee in Texas won a workmen's compensation case against an employer whose safety bingo program discouraged employees from accurately reporting incidents.[1] Depending on other aspects of a company's loss prevention efforts, any incentive that encourages employees not to report incidents may increase a company's liability should an incident occur. Such an incentive certainly increases an employee's ability to successfully file a compensation claim well after the typical time periods established for such a claim.

Regardless of the kind of award system you design, you will have to create an internal marketing campaign to promote your safety effort with employees. You will want to consider posters, announcements in safety meetings, articles in

---

[1] *Paragon Hotel Corporation v. Ramirez*, 783 Southwest 2nd 654, 1990, El Paso Court of Appeals.

newsletters, and other methods of promoting and communicating the new process. Whether you choose to announce the award program is a separate question. As discussed below, sometimes you may be better off not to "dangle the carrot" by kicking off your award program with a great deal of hoopla.

In planning a safety awards and recognition program, be aware that many people have had negative experiences with traditional awards programs. Employees may have had experience with awards that were not meaningful, favoritism, programs that did not fairly recognize their efforts, or awards that were determined by chance rather than performance. Such experiences often create resistance to well-designed award and recognition programs. Employees who resist awards programs, however, are normally receptive to team celebrations of success and recognition for individual efforts and contributions.

The administration of recognition and celebrations is as important, perhaps even more important, than the initial planning. Chapter 22 will provide additional details related to using recognition and celebrations as effective positive consequences.

11.1 OVERVIEW OF SAFETY AWARDS AND INCENTIVES

Safety award programs are usually fairly easy to create, but an effective incentive program is relatively complex and involved. Safety award programs are most effective if they are administered locally with a high level of involvement by the safety teams and line management. Incentive programs, on the other hand, usually require a higher level of approval and are often administered organizationwide.

Figure 11.1 shows various options for safety awards. Obviously, you may choose to combine several of the options. For example, you might combine a concurrent safety award program, a recognition process, and one of the compensation options. Your design team can create a recognition and concurrent safety award process for your steering committee to implement at the local level, but changes to compensation will usually require working through executive management.

11.2 SAFETY RECOGNITION

The least formal safety award process tries to maximize use of personal recognition, often in the form of appreciative feedback. Such recognition may be provided in safety meetings or as part of other group meetings or activities or it may be provided privately to individual employees. With this approach, your design team might recommend that steering committees and management put recognition on the agenda at the beginning of every meeting and allow time for participants to pass along their thanks to others in the organization who have done something worthy of recognition. You might also promote the use of written thank-you notes, perhaps using copies on bulletin boards to publicize both the appreciation and the specific contribution. This approach is usually more successful when it is supported through a training process that helps everyone understand the rationale for the approach and

	Safety Recognition	Simple or Concurrent Safety Awards	Tiered Safety Awards	Safety Integrated with Traditional Compensation	Safety Incentives
Recognition or Awards	Social recognition, often a formal agenda item in meetings and written Thank-You notes	Fixed award for everyone - OR - Menu of items or events	Multiple tiers with menu of recognition items or events appropriate to each tier	Merit increase or annual performance bonus	Incentive bonus or time off provided upon meeting criteria
Criteria*	Usually not established in advance	Specified in advance for each separate program	Provides guidelines on behaviors or results appropriate for each tier	a) Number of observations b) Lost-time injury rates c) Worker s compensation cost	a) Recordable injury rates b) Lost-time injury rates c) Worker s compensation cost
Participants	Individuals and teams	Individuals and teams	Individuals and teams	Usually management personnel	Individuals and teams
Considerations	Often difficult to ensure that all levels and functions are included equally	Simple programs must be constantly varied to maintain novelty and keep awards meaningful Multiple, concurrent awards provide better distribution and variety of awards and also increase the probability of an employee receiving an award	Provide good variety and wide distribution of awards Increase the probability of awards for those who champion safety	Requires that those who evaluate personnel know the applicable safety activities	Usually requires management or staff observations of safety activities in their work areas Gainsharing often funded by savings in worker's compensation

* Ideally, awards are based on behavioral criteria, not simply a fixed period of time without an incident.

Figure 11.1. Design options for safety awards and incentives.

how to provide effective personal recognition. Often, this can be part of the observer training provided to all employees, supervisors, and managers.

You should encourage the steering committee to arrange for some form of recognition and appreciative feedback for your observers. In particular, observers who are conscientious about completing their observations should receive meaningful recognition from their supervisors and managers. This recognition might be a simple, personal expression of appreciation during safety meetings or be made via thank-you notes or letters of commendation. Ultimately, providing some form of recognition for completing observations will be important to maintaining the observation process.

Appreciative feedback is a very important element of the safety award options discussed below. All awards should include a clear, pinpointed statement of what was done to earn the award and a suitable statement of appreciation.

11.3 SIMPLE AND CONCURRENT SAFETY AWARDS

The easiest option for creating a simple safety award program is to identify the criteria and a potential award or recognition event or perhaps a menu of awards and recognition events appropriate to the criteria. The idea is to strive for planned, yet spontaneous recognition. It is planned in that you know what pinpointed behavior or accomplishment you plan to reward, but it is spontaneous in that you do not know when the criteria will be met. These award programs resemble traditional safety awards except that employees earn the awards based on specific actions that promote safety, not by going a fixed period of time without an incident.

In contrast to most traditional safety awards, simple awards can support on-the-job safety when provided to employees for simply going without an incident. To be effective, the awards must be small enough to not impact on reporting and be based on avoidance of injuries for a relatively short period of time such as a month. Bill Hopkins and Jim Getting have described the use of trading stamps as an incentive to reduce incidents in mining (see Chapter 28). This study shows the effectiveness of providing awards to employees who complete a month without injury in conjunction with other safety efforts. Such programs can work especially well in combination with an effective observation process.

Typically, the design team creates a simple award program and suggests several possible criteria and potential awards, then allows the steering committee to modify the plans to best support the behavioral process once employees begin doing observations.

If the safety award is unchanging and the same for everyone (such as jackets, caps, coffee cups, or other such items), the steering committees should realize that they are usually creating an award *program*, not a process that can continue on indefinitely. As such, it should have a planned life cycle. The steering committees should maintain such a program for some fixed period, such as six months; then they must design and initiate a new award program to take its place. By varying the awards and criteria, the steering committees keep the awards novel and help keep employees interested and thinking about safety.

In order to create a longer lasting recognition *process*, your design team may suggest that the steering committees create a flexible award process based on a menu of awards rather than a simple fixed award. By using a menu of awards, your steering committee can select an award that can be more meaningful to a given individual or team. Some of the more elaborate procedures described in the following section on tiered awards may be useful in developing a menu of reinforcers for safety awards.

In addition, your design team may choose to encourage the steering committee to start with a simple award process and add additional concurrent awards later.

	Individual Safety Recognition	Team Safety Celebrations
Administrated by	Steering committee	Steering committee

Criteria	Completing scheduled observations
	Safe practices (from checklists)
	Quality of a safety meeting
	Specific contribution or level of support
	<u>Identifying</u> a significant unsafe practice or condition
	Near-miss report
	Off-the-job or home safety
		Targets set by steering committee (monthly or quarterly)
		
		Suggested criteria:
		   Number or percent of observations
		
		Other criteria (usually downplayed)
		   Percent safe
		   Percent safe over time
		   Number of unsafe acts reported

Award or celebration	Rub-off lottery tickets • Each area gets a number of cards equal to the number of its employees • 25% of all tickets are winners • All tickets good for drawing • Menu of awards selected by steering committee • Maximum value $10 • May include gift certificates	BBQ prepared by area manager Breakfast Chicken lunch Cookouts Dinner Donuts Ice cream social Pizza Watermelon break
Other elements	Employees may earn recognition for activities and contributions not on initial list of criteria Recognition usually provided as part of a meeting, often in conjunction with team celebration	Not announced in advance May be used as opportunity to provide individual recognition

Figure 11.2. Example of concurrent safety award processes providing individual recognition and team celebrations.

108 STEP 4: DEVELOPING RECOGNITION AND CELEBRATION PLANS

This approach eventually leads to a more complex, concurrent safety award process. For example, the initial safety award program might provide a catered meal for a group of employees who achieve their percent safe target for proper lifting in their area. The steering committee might also create an additional awards program for groups or shifts of employees who maintain 90 percent or better of their scheduled observations, then perhaps several months later add another award program for safety suggestions.

Another example of a concurrent safety awards process would involve the creation of one plan for recognizing individual contributions and another plan for celebrating team successes. Figure 11.2 presents an example of such a concurrent awards process.

11.4 TIERED SAFETY AWARDS

The best safety award programs have multiple types of awards appropriate for different kinds and levels of performance. They are similar to frequent-flyer awards: You get the awards no matter when or where you fly, and the more you fly, the greater the value of the award. Multiple tiers establish different levels of contributions with corresponding awards. Such an approach often builds on or incorporates simple awards programs and the use of personal recognition. It may expand simple or concurrent award processes by combining them into a single award process in which the concurrent awards are placed within the appropriate tier. The safety recognition process may become the lowest tier of the process, often with additional guidelines regarding the types of behaviors that would justify recognition. The idea is to recognize a large variety of both individual and group contributions.

The design of a model tiered safety awards process often requires completion of several tasks, as shown in Table 11.1.

Develop the criteria for different levels of performance. Figure 11.3 provides a worksheet for designing the tiers for this type of safety award process. One of the key elements is that the guidelines or criteria for awards are clearly established, although they may be adjusted if a safety-related behavior occurs that deserves

TABLE 11.1. Tasks That Must Be Completed in Design of a Tiered Safety Awards Process

Task Number	Activity
1	Develop the criteria for different levels of performance.
2	Identify potential awards and forms of recognition.
3	Survey area personnel regarding their preferences.
4	Finalize a menu of celebrations and awards for each level of performance.
5	Plan the award delivery process.
6	Present the safety award plan to management.

RECOGNITION WORKSHEET

Level	Criteria	Menu of Recognition Activities or Awards
1		
2		
3		
4		
5		

Figure 11.3. Sample worksheet for planning a tiered safety award process.

recognition. The idea is to create a flexible system with guidelines that assure consistency. Plan to establish three to five levels of recognition for both individual contributions (such as submitting a high level of implemented safety suggestions) and team successes (such as achieving a safety goal).

As your first task, identify the kinds of activities and accomplishments you wish to recognize. You may want to brainstorm an initial list of tentative criteria, then ask several design team members to use this list to develop a more complete list of safety contributions that you might want to recognize and sort them into different levels. Figure 11.4 provides an example of possible criteria that might be appropriate for five levels.

RECOGNITION WORKSHEET

Level	Criteria	Menu of Recognition Activities or Awards
1	Individual who submits safety suggestion that is approved for implementation. Individual who leads safety meeting. Individuals who participate on steering committee.	
2	Individual completes all scheduled observations for a month. Individual with most valuable safety suggestion in a given month. Safety team for achieving milestones.	
3	Area achieves new high on safety index (from observations). Individuals who complete all scheduled observations for a quarter. Area that meets its safety target for a month.	
4	Area meets its safety target for a quarter. Area that maintains safety observation process for a year.	
5	Area that maintains safety observation process for a year and betters industry average for injuries.	

Figure 11.4. Example of a recognition worksheet showing possible criteria.

STEP 4: DEVELOPING RECOGNITION AND CELEBRATION PLANS

Identify potential awards and forms of recognition. Your next task is to develop a list of possible recognition and award ideas. Before establishing the award system, you should find out how much money is available for steering committees to spend on safety awards. Once the award system is established, awards should be a standard item in the annual budget.

You should keep each award relatively small, generally $10 to $25, although the rare awards in your highest level might be as much as $250. The idea of an award program is for the company to express its appreciation in a way that is meaningful to the employee(s), but it is not supposed to be compensation. Also, you do not want to place too significant an award on safety because of the danger of damaging

Social		Letter from supervision
		Letter from upper management
		Name of individual or team on bulletin board at plant entrance
		Recognition memo from managers from other areas
		Write-up in plant publication
		Letter to spouse or family
Work related		Plant tour
		Special assignments
		Opportunity to participate on steering committee
		Training for different job within area
		Other training courses
		Time to discuss issues with supervisor
		Assignment to preferred duties
		Area discussion with plant manager
		Opportunity to participate in management presentation
Tangible	$25 – $50:	Dinner
		Gift certificate
		Pizza for team
		U.S. Savings Bond
		Walkie-talkie
	$50 – $100:	Car stereo
		Citizens band radio
		Gift certificate
		Radio and tape player
		U.S. Savings Bond
	$100 – $200:	Camping outfit
		Compact disc player
		Compact disc player (auto)
		Color television
		Gas edger
		Gift certificate
		Small television set
		U.S. Savings Bond

Figure 11.5. List of possible safety recognition activities or awards.

the integrity of the reporting and observation processes. If the awards are overly valuable, they can negatively influence these processes.

Again you may simply brainstorm recognition awards and events with the design team. Figure 11.5 presents a list of possible safety award items and events that your team may want to consider. You may also want to solicit additional ideas from employees outside the design team (e.g., using the form in Fig. 11.6 during the kickoff meeting to get such input).

Area Safety Awards Survey

Your steering committee is working on a team management process to enhance our safety efforts. As part of this process, we want to create celebrations of our safety successes that are fun for everyone involved.

Please assist us by responding to the following survey items. We are looking for awards and activities to recognize both individuals and teams. Your input into our safety recognition activities will help us develop a more effective program!

1. For each of the following approximate dollar amounts, what are some awards you or your team would enjoy?

    $200 (for example, VCR): _____

    $100 (for example, portable radio-tape player): _____

    $50 (for example, radio or dinner for two): _____

    $25 (for example, less expensive dinner for two): _____

2. What suggestions do you have for creating a fun safety award process? Use the space below and on the back of this sheet (if you need additional space) to make your comments and suggestions.

Figure 11.6. Sample initial survey to get ideas for recognition menu.

SAFETY RECOGNITION SURVEY

Name:_____
(Optional)

Please help the steering committee by giving us your input. Circle the choice that indicates how important each of these forms of recognition is to you.

1. Social

    1.1 Letter from supervision

        Low Importance   Somewhat Important   High Importance

    1.2 Letter from upper management

        Low Importance   Somewhat Important   High Importance

    1.3 Name of individual or team on board at plant entrance

        Low Importance   Somewhat Important   High Importance

    1.4 Recognition memo from other department managers

        Low Importance   Somewhat Important   High Importance

    1.5 Write-up and picture in plant publication

        Low Importance   Somewhat Important   High Importance

    1.6 Letter to spouse or family on safety achievement

        Low Importance   Somewhat Important   High Importance

Figure 11.7. Sample survey to help rank value of items for recognition menu.

Survey area personnel regarding their preferences.[2] The reason for establishing an awards menu is that it enables you to select an award or form of recognition that is meaningful to the recipient. You will therefore find it useful to have some idea of how significant the awards are to other employees. You could simply work with the

---

[2] This approach is based on "A Reinforcer Profile in 8 Easy Steps," *Performance Management Magazine,* 9(2), 3–7 (1991).

Form of Recognition	Preference Score*
Dinner at restaurant of your choice	4.8
Catered barbecue	4.6
Assignment to preferred duties	4.3
Time off	4.3
Gift certificate	4.3
Lunch certificate	4.0
Time to discuss issues with supervision	3.9
Opportunity to participate in steering committee	3.7
Area discussion with plant manager	3.7
Letter from upper management	3.4
Training to do a different job within area	3.2
Donuts and coffee	3.2
Longer break	3.2
Special assignment	3.1
Name of individual or team on board at plant entrance	3.0
Write-up and picture in plant newsletter	3.0
Pizza party	3.0
Barbecue cooked by supervisor or management	3.0
Letter from supervision	2.8
Breakfast cooked by supervisor	2.8
Breakfast cooked by manager	2.8
Letter to spouse or family on safety achievement	2.6
Recognition memo from managers from other areas	2.6
Opportunity to participate in management presentation	2.6
Items with company logo	2.6
Plant tour	2.3
Plaques	2.3

* The preference score is the average of all survey responses scored as 1, 3, or 5, with 5 being the "High Importance" response.

Figure 11.8. Sample data on employee preferences, ranked and grouped into tiers based on recognition survey.

design team to rank the items on your list of potential recognition items and events. Or you may want to use a survey. Figure 11.7 shows a sample survey that you might construct based on your initial list of award ideas to help determine how significant those awards are to employees. You may choose to survey other design team members or a broader sample of employees.

When you get the surveys back, summarize the data and develop a prioritized list based on the survey responses. Figure 11.8 presents an example of a prioritized list of recognition items showing the average scores based on survey responses. Notice how the list is divided into five categories based on logical breaks in the preference scores. At the end of this task, you should have a list of individual and group awards and recognition ideas that are ranked within tiers that correspond to the number of criteria levels you identified earlier.

Finalize a menu of celebrations and awards for each level of performance. The remaining task in designing your safety award process is to match the performance criteria and the award levels. You then have a complete model for steering

RECOGNITION WORKSHEET

Level	Criteria	Menu of Recognition Activities or Awards
1	Individual who submits safety suggestion that is approved for implementation. Individual who leads safety meeting. Individuals who participate on steering committee.	Letter to spouse or family about safety contribution Opportunity to participate in management presentation
2	Individual completes all scheduled observations for a month. Individual with most valuable safety suggestion in a given month. Safety team for achieving milestones.	Breakfast cooked by supervisor or manager Letter from supervisor Barbecue cooked by supervisor or management Pizza party Write-up in plant newsletter
3	Area achieves new high on safety index (from observations). Individuals who complete all scheduled observations for a quarter. Area that meets its safety target for a month.	Special assignment Longer break Donuts and coffee Opportunity for additional training Letter from upper management
4	Area meets its safety target for a quarter. Area that maintains safety observation process for a year.	Area discussion with plant manager Opportunity to participate on steering committee Time to discuss issues with supervision Lunch certificates
5	Area that maintains safety observation process for a year and betters industry average for injuries. Individuals who complete high percentage of observations for the year. Individuals who makes year s best safety suggestion.	Gift certificate Time off Assignment to preferred duties Catered barbecue Dinner with spouse at restaurant of choice

Figure 11.9. Sample of a completed recognition planning worksheet that provides criteria and menu of possible forms of recognition.

committees to use in developing a safety award and recognition process. Figure 11.9 presents a completed safety award planning worksheet that lists criteria for both individuals and teams and awards appropriate for both at each level.

Plan the award delivery process. The delivery of the safety awards is critically important to the overall success of your award process. Your goal is to create a process that is fun for everyone in the area and a true celebration of safety achievements.

The delivery process will vary for each level of award, and individual recognition will often differ from group celebrations. Generally, group celebrations are a good time to recognize individual contributions. In this way, everyone wins—the employees enjoy the group celebration and individuals get attention for their contributions.

The important task here is not planning all of the details in advance for each award on your menu but rather suggesting an award process that the steering committees can administer and the criteria that will trigger an award and celebration. You will probably want the steering committees to oversee the process and establish a regular cycle of review to identify who has earned a recognition or safety award. Planning safety recognition and celebrations should be a regular item on the steering committee's agenda.

Ideally, your procedure will be flexible enough so that each award is selected and planned in a way that is significant to the person or group that earned it. As stated earlier, the reason for a menu is to allow you to select something meaningful for significant safety contributions. You will usually want to ensure management's involvement in delivering awards and a level of management appropriate to the level of the awards. Such personal involvement by management helps ensure the credibility of the awards process.

In addition, the menu of celebrations and awards should only serve as a guide. Your steering committee should be encouraged to identify meaningful awards and celebrations that are not on the menu but in approximately the same category of recognition items or events appropriate to a particular contribution. As you select new safety awards, these should be added to the menu so that they can be considered in the future.

Although lotteries and contests can be a fun and positive way to provide additional awards, they sometimes create problems because of the competition and danger of hurt feelings. Some contests can be effective, however, especially in some of the support programs discussed in Chapter 23. Figure 11.10 compares the features of contests, recognition programs, and incentive compensation plans.

Lotteries can also be used effectively to deliver safety awards and increase the value of the more significant awards. The key is to make delivery a fun celebration for everyone and to have many small awards in addition to a main prize so that you have many "winners." Lotteries can be especially effective as part of the celebration during a safety awareness day (see Chapter 12).

Lotteries are likewise effective in conjunction with token systems in which observers dispense tokens to employees as part of the observation process. Typically the token is a "thank-you for being safe" card that is also a lottery ticket. Such tickets are handed out, for example, when an area scores 100 percent

116 STEP 4: DEVELOPING RECOGNITION AND CELEBRATION PLANS

Contest	Recognition	Incentive Plans
Announced	May be either announced or unannounced	Announced
Based on behavior and/or results	Based on behavior and/or results	Based primarily on results
Criteria are typically competitive	Criteria are flexible and may include improvement	Criteria are fixed
Provide motivation only to top performers	Provide motivation to all employees	Provide motivation to all participating employees
Design is very easy	Design is fairly easy	Design is more difficult

Figure 11.10. Comparison of contests, recognition, and incentives.

safe on an observation checklist. Employees can continue to collect the lottery tickets up to a deadline date, perhaps the safety awareness day cited above. The more tokens an employee has, the greater that employee's chances of winning. Employees might earn tokens in other ways as well, as by submitting safety suggestions or participating in a slogan contest. (Note too, however, that a lost-workday case during this time prevents the area's participation in the lottery or may even end the lottery altogether. No one wants to celebrate if someone has been injured.)

The advantage of token systems is the immediate, tangible presentation of items that provide direct support for on-the-job safety compliance. The tokens provide both positive feedback and a positive source of motivation for compliance and participation. Again, using an award that is very significant or has great monetary value creates problems because of the disappointment and hard feelings created by those who do not win the lottery. Done correctly, token systems can be fun and help build pride in an organization's safety efforts.

Your records of safety awards should include the names of employees who earned awards, the level of award, and the event that celebrated their achievement. Such records will enable you to check what was done in the past and prevent the possible embarrassment of giving the same award to the same individual or group.

Present the safety award plan to management. If you are using periodic planning meetings, completing your safety award plan is another of the milestones in your project plan. Before initiating the award process, you should get management's input on and approval of your criteria, awards, plans for delivering the awards, and, perhaps most importantly, the budget that you will need to support the planned process. Generally, you will have to make some assumptions about how many individuals will earn recognition and how many teams will earn celebrations so that you can estimate the likely costs of your process. At this stage, you generally are better off slightly overestimating what you will need to fund this part of your process rather than risk having to ask for more money later or, worse, having co-workers earn awards that you cannot provide because you have exceeded your budget! Certainly, when you present your budget for approval, you will want to outline the assumptions you made so that management understands the basis for your request. As mentioned earlier, you should also carefully outline management's role in the delivery of safety awards, especially for awards that require managers to write commendation letters—or cook breakfast on the third shift! Also, you will usually need to prepare a budget for your recognition plan. Once management approves this budget, the steering committee must administer its recognition and celebration plans within the constraints of their budget.

11.5 SUPPORT THROUGH TRADITIONAL COMPENSATION

Compensation can be an effective component in maintaining an effective safety management process. This approach sometimes applies only to managers and supervisors responsible for maintaining the behavioral safety process. How you link safety to salary increases depends on how such increases are determined within your organization. It may include adding safety as an explicit dimension of the appraisal form or building in a process that allows managers to specify annual goals for maintaining the safety process, such as a specific percentage of completed observations each month. As with safety awards, the emphasis should be on maintaining the safety system. However, compensation decisions will often include an assessment of the area's safety performance, including both the safety process and the resulting incident rates. The assessment process should balance evaluation of both process and outcomes.

11.6 SAFETY INCENTIVE COMPENSATION

As with other forms of tangible rewards, the challenge is to support the observation and feedback processes without generating false reporting. In general, the use of bonuses tied to safety is not recommended, except when safety is included as part of an overall management scorecard.

Effective safety incentive compensation typically includes three elements:

1. Financial incentives based on measured safety outcomes that reflect economically on the company's success
2. Recognition and feedback that support maintaining the behavioral safety process and employees' compliance with safety procedures
3. Severe penalties, including usually loss of employment, for falsely reporting either incident or observation data

When setting up a cash incentive for safety, the two rules of thumb from the beginning of the chapter still apply: The award should be relatively small, and it should be based primarily on maintaining the observation system and its components.

Cash awards and paid time off are both such significant awards and may be considered forms of compensation. You must use them carefully to ensure that they support your safety process and do not encourage false reporting.

If you plan on developing an incentive compensation plan, it should be based on outcome measures such as total workmen's compensation claims, recordable injuries, or lost-workday cases.

Total Workmen's Compensation Claims. The total dollars spent on compensating injured workers would seem to be an ideal basis for an incentive bonus. This measure has several advantages and avoids the disadvantages of incentives based on working a fixed period without injury. On the plus side, it is easy for both managers and employees to understand. It is also an important cost that directly impacts a company's profitability. Because it affects profits, it is often easy to sell to upper management. Documented savings in compensation costs could be used to fund the safety portion of the incentive bonus. This approach is particularly good for companies looking for ways to partner with employees to share both the risks and the potential savings associated with safety. A complete implementation of the behavioral safety process allows companies to make employees responsible for the safety process and share the economic benefits of good safety performance. If the safety efforts do not produce savings in workmen's compensation costs, the incentive system has no money to fund bonuses. The savings may also fund a separate bonus pool for managers and employees.

Using total dollars spent on compensating injured workers is an appropriate incentive criterion. First, it does not directly discourage reporting incidents. Both managers and employees have very little incentive to try to hide minor incidents because such incidents add very little to the total compensation costs. Second, it encourages managers to be proactive in managing an injured employee's return to work. The sooner an employee gets back to work, the less it costs the company.

However, this measure has several significant problems. First, it is often not readily available or is only available several months after the fact. Second, the most significant costs often accrue over a long period of time and may not be evident in the initial months following the incident. These problems make using a 12-month moving average a good basis for bonus calculations, but even 12 months can be too short a time frame to assess the total costs associated with a serious injury.

Lost-Workday Cases. Lost-workday cases can also be used effectively as the criteria for safety incentive systems. Managers and employees are not likely to hide serious injuries, especially if they understand that to do so would likely cost them their jobs. Payouts are generally funded through a pool set aside for the costs associated with serious injuries. A large food processing chain, for example, charges each of its locations $500 that goes into a pool to cover such costs. At the end of the year, unspent funds are used for bonuses for managers at sites that have had no lost-time injuries.

OSHA-Recordable Cases. Generally, OSHA recordables should not be used as a stand-alone basis for incentive bonuses. The problem, as we have talked about before, is the difficulty of ensuring the integrity of reporting these injuries. While OSHA-recordable incidents or incident rates are not a good stand-alone basis for an incentive bonus, they may be part of the criteria, as discussed in the next paragraph.

Safety Index and Score Card Approach. The best approach for creating a safety bonus is probably to use an index that combines and weights several measures. Lost-time injuries or compensation costs might be weighted heavily, while recordable injuries and even process measures can be included and weighted less. The advantage of this approach is that it balances the emphasis on process and outcomes, and recordable injuries can be included but given a small enough weight that the integrity of reporting is easier to maintain.

11.7 GENERAL GUIDELINES ON SUPPORTING SAFETY MOTIVATION

As a review of the options for using recognition, awards, and incentives, Figure 11.11 summarizes the methods of providing motivational support for safety.

Rules for Effective Safety Motivation	Considerations
Do provide recognition for employees that appropriately complete their observations.	Such recognition is important to maintaining the observations.
Do make recognition personal and meaningful to recipients.	Steering committees and managers should plan to make the recognition meaningful.
Do not use data from employee observations as the basis for salary evaluations or disciplinary action against employees.	Managers and supervisors should use documented observations made outside of the behavioral safety process as the basis for decisions regarding salary and disciplinary actions.
Do evaluate supervisors on the percent or number of observations completed in their work areas.	Use this as one indicator to ensure the process maintains. Be careful that supervisors do not pressure employees to participate.
Do evaluate supervisors on the basis of management observations conducted in each work area.	Such observations should include discussions with employees about how the supervisors are managing the behavioral safety process.
Do not evaluate supervisors on the basis of the % safe or safety index from their work areas.	The impact of such evaluations would potentially distort the observation data.

Figure 11.11. Considerations for supporting appropriate motivation for the behavioral safety process.

12 Step 5: Planning Training and Kickoff Meetings

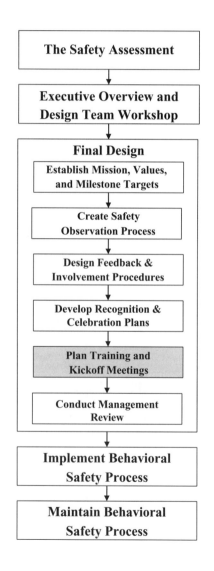

Once you have planned recognition and celebrations to support your safety process, you are ready to plan how you will introduce the process to employees, train observers, and other training that may be important to the success of your process. One of the decisions that you will have to make is whether you will need to have a kickoff meeting (or meetings) to introduce your process. Generally, you will need one unless you have a fairly small site and can get all employees through observer training in a fairly short time.

12.1 OBSERVER TRAINING

In general, you should require that all employees complete observer training, even if you have a voluntary process and employees say they will not volunteer. The reason is that observer training provides a much higher level of understanding of your process than you can achieve in a kickoff meeting. This increased understanding will contribute to greater acceptance and support even from employees who do not volunteer to conduct observations. In addition, by requiring all employees to complete the training, you ensure that employees are making an informed decision; that is, they know exactly what they are volunteering for or opting out of.

The observer-training workshop will usually require 8 hours. Figure 12.1 presents a typical agenda for observer training. During the first hour of the workshop, employees learn the rationale for the behavioral approach and how the checklists were created. The remainder of the day is spent learning how to pinpoint behavior (see Chapter 21), how to conduct observations, and how to discuss their observations with their associates. When learning to conduct observations, employees need to practice conducting observations and discuss what they observe. To aid in this practice, you should videotape several short work samples that instructors can use to train workshop participants in how to use the observation checklists. Generally, actual, unstaged work samples work best for this purpose and

---

Three objectives:
- Understand and support the process
- Be able to use the checklist
- Be able to discuss observations effectively

Overview and introduction
- The design team
- Why a behavioral approach
- What is a behavioral safety process

The safety checklist(s)

Observation skills and practice

Feedback skills and practice

Figure 12.1. Typical agenda for the observer-training workshop.

provide realistic opportunities for practice observations. After the practice observations and a discussion of how to complete the checklists, participants should break up into pairs and role play how they would provide feedback to the employee they just observed in the videotape. After the first role-playing exercise, participants should switch roles so that both have the opportunity to practice the feedback discussion. If you can arrange it, after completing several exercises of this type, you should allow the employees to go out into the workplace and conduct an observation, then return to the classroom to discuss their experience. In this way, the observers get significant practice conducting observations, which will ensure they have the skills to participate in the process.

12.2 PLAN KICKOFF MEETING(S)

If you are not able to train all employees in your facility in a fairly short time, you will need to conduct kickoff meetings with employees to introduce the behavioral safety process. The design team should provide a suggested meeting agenda and make general recommendations for the meetings, usually conducted by the steering committee. Covering all work areas and shifts will generally require several meetings. Small groups of 8 to 10 individuals are ideal as they provide participants with a better opportunity for questions and discussion than do larger groups. Often the agenda will look very much like the presentation to management described in the next chapter.

Depending on the size of your organization, design team members should participate in as many kickoff meetings as possible to enable employees to ask questions of those who designed the process. You may also want to suggest a role for management in the kickoff meetings. For example, you might suggest that the

---

Purpose: Introduce behavioral safety process to employees

- Background on how team came into being
- Objectives of the observation system
- Process for developing observation system
  - List of initial pinpoints
  - Final checklist(s)
- Observation procedure
- Management's role
- Next steps
  - Observations begin
- Questions and discussion

Figure 12.2. Typical agenda for kickoff meetings to introduce the observation system to employees.

steering committee invite either a representative of upper management or the area manager to comment on management's support for the safety improvement efforts.

Figure 12.2 presents a typical agenda for a kickoff meeting. As with the other elements of the process, if your design team is developing a process that will be initiated by the steering committee, the design team should provide a suggested agenda. It will enable each steering committee to develop its own kickoff meeting agenda based on the model provided by the design team.

12.3 PLAN TRAINING NEEDED TO SUPPORT THE PROCESS

As you consider implementation of the observation and feedback processes, you will need to consider the existing skills and training needs of those you want to participate. For the behavioral safety process to be successful, employees will need skills in the four areas given in Figure 12.3.

Who	Description	Specific Skills
Management	How to support the implementation and maintenance efforts of steering committee	How to participate and follow up on implementation efforts How to use data on completed observations to evaluate the process Observation and feedback skills described above (so that they can model the process)
Steering committee members	How to implement and maintain the behavioral safety process	How to use the guidelines provided by the design team to: • Refine observation checklists • Conduct the area kickoff meeting • Initiate observations and provide feedback • Train others to conduct observations and provide feedback • Initiate a recognition program in support of the process. How to evaluate and problem solve the process
Observers	Observation and feedback skills	How to use observation checklist How to provide feedback on observations
Steering committee members and observers	Leading safety meetings	How to discuss data How to set improvement targets
Employees who conduct the specific job tasks	Job-specific safety skills	Specific practices identified from the safety observation checklist

Figure 12.3. Skills needed to support behavioral safety process.

124 STEP 5: PLANNING TRAINING AND KICKOFF MEETINGS

If your design team is planning a behavioral safety process to be initiated by the steering committee, you will need to ensure a well-developed formal training process for steering committee members. The primary objective of this training is to teach steering committee members to implement and support the behavioral safety process. Implementation training for the steering committee is often the most complex training involved in the process (other than training for the design team). Steering committee members will need to know how to take the guidelines developed by the design team, tailor those plans for their respective work areas, and then initiate and maintain each element of the process. Such training usually requires a formal workshop that resembles the initial training provided for the design team.

Steering committee members who will be implementing the behavioral safety process will also benefit from an understanding of the rationale or basic theory underlying the behavioral safety process, including the observation and feedback procedures. They will provide better and more consistent support for the elements of the behavioral safety process when they understand the reasons for those elements.

In addition to identifying the training needed to make implementation successful, your design team should decide on the most effective way to deliver each type of training. You want a training process than balances effectiveness with minimal cost and disruption to the workplace. The training options include the following:

- Individual coaching (tell, show, observe, and provide feedback)
- Mentors
- Seminars or workshops
- Videos or slides

While you will usually want to provide a detailed workshop for the steering committee and line management, you may consider individual coaching and mentors for training new observers. Such an on-the-job training process is a less disruptive process than providing workshops or seminars. The steering committee may later arrange additional training to provide an understanding of the rationale for the behavioral safety process as part of ongoing safety meetings.

To address job-specific skills, consider allowing employees to make training videos or a slide show. A video of actual work is an excellent tool for training observers in how to use the observation checklist and can also be used to illustrate the use of the checklist during area kickoff meetings. Employees might also develop training materials showing near-miss incidents or past incidents. Such a strategy provides an effective training tool that creates a high level of involvement. (See the Chapter 14 for an additional discussion of such training and involvement efforts.)

13 Step 6: Conducting Management Review

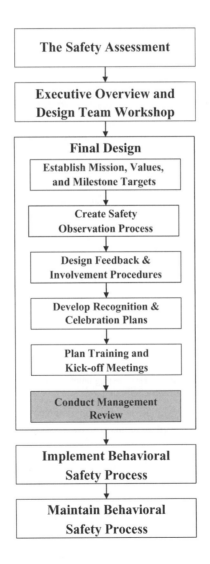

STEP 6: CONDUCTING MANAGEMENT REVIEW

Once you have completed your planning efforts, the design team should present the behavioral safety process to management for its input and review. This step should be a cricital milestone on your initial schedule. This meeting is an opportunity for management to

- provide input to the safety process,
- approve implementation of the new process,
- commit to its personal involvement,
- approve the budget for recognition and celebrations, and
- provide recognition of the efforts of the design team.

Getting management's input and suggestions for your observation system will help ensure that it will support your implementation efforts. This will also provide an opportunity for ensuring that management understands each component of the behavioral safety process at each stage of the implementation. Figure 13.1 presents a typical agenda for such a meeting. As stated before, this agenda is appropriate for management review for a process that was designed relatively quickly using two-day planning meetings. A longer planning process in which the design team meets for 2 hours every other week should have a milestone review at the completion of each significant stage of the planning.

All design team members should participate in the presentation. Each design team member should take responsibility for presenting one agenda item, with all members participating in a question-and-answer session at the end. Managers will

---

Purpose: Get management's input and approval for the design team's plans

- Introduce design team members
- Overview of the design process
- Team's mission and values
- Overview of the observation process
  - Observation checklist
  - Steps in conducting an observation
  - Feedback discussions
- Observation checklist and procedure
- Data collection and analysis
- Rollout plans
  - Kick off meetings
  - Observer training
- Individual recognition and team celebrations
- Conclusions and request for support

Figure 13.1. Typical agenda for the management presentation.

usually be particularly interested in the observation checklist and how it is constructed. They will often have questions and may have additions that your team should consider. You should also pay special attention to the discussion of recognition and celebrations. In the presentation, you will have to explain the assumptions that you made in creating the budget and explain how you calculated your anticipated costs. Also, be clear about your request for funds as well as the other forms of support that you want from management.

Design team members will be nervous about this kind of presentation. For many it may well be the first time they have given a formal presentation of this type to management. Try to have a dry run so that team members can practice their presentation and use of visual aids. Generally, they should have slides or a computer presentation to support their message. Pay special attention to the transition from one team member to the next. In the end, the presentation does not have to be a polished professional presentation that you might expect from consultants or experienced managers. The team's enthusiasm and sincerity will make the presentation work. Management will be able to tell that the team members are "speaking from the heart" and *that* will make the presentation effective.

Management's role should be to understand the process and reinforce design team members for their participation in planning the behavioral safety process. They should ask questions, being careful to be positive and supportive of the team's efforts. Obviously, managers need to voice any concerns they have, but they should keep the overall tone of the meeting positive and support the design team's enthusiasm.

14 Implementing Behavioral Safety Process

The implementation of a behavioral safety process usually will be the responsibility of your steering committee. The implementation process outlined in Table 14.1 provides the team with the opportunity to change the checklist to meet particular needs of each area or site. Such additional input will not be necessary if your design team and safety team are the same group.

TABLE 14.1. Implementation of Behavioral Safety Process

If:	Then:
Your design team conducts periodic planning meetings and is implementing the behavioral safety process concurrently with their planning efforts.	Read this chapter as it relates to the implementation of the particular element on which you are currently working.
Your design team is implementing in a pilot area.	Continue with this chapter to clarify the responsibilities of steering committees and management for implementing the behavioral safety process within the pilot areas.
Your design team planned the behavioral safety process and will become the steering committee and take responsibility for implementation.	As you read this chapter, substitute *design team* for *steering committee*. Continue with this chapter to review the respective responsibilities of the design team and management for implementing the behavioral safety.
Your design team planned the behavioral safety process but other steering committees will be responsible for implementation.	Continue with this chapter to clarify the respective responsibilities of safety teams and management for implementing the behavioral safety process throughout the organization.

14.1 CONDUCT TRAINING FOR STEERING COMMITTEES

If your steering committee includes employees that were not on the design team, then the first step in initiating the behavioral safety process is to train new steering committee members. This training typically requires a two- or three-day course created by the design team. Ensure that all levels of management participate and learn how to conduct observations and deliver feedback so that they can model the process when they visit the work areas (if you have planned management participation as part of your effort). The training should clarify the roles outlined below for the process owner, the steering committee, and management. The steering committee's role is discussed in greater detail in Chapters 20 and 22.

14.2 ESTABLISH A PROCESS OWNER

As you move into implementation, one member of each steering committee should be appointed as the local *process owner*, or coordinator, for your new safety process. This person may be a staff member, supervisor, or hourly employee who will serve as the employee safety coordinator. If the position is assigned to an hourly employee, you may want to rotate it annually.

The behavioral safety coordinator has primary responsibility for ensuring that the steering committee implements and maintains the behavioral safety process. The duties include working with management to develop agendas for steering committee meetings, scheduling observations, reviewing completed observation checklists, and all other tasks not explicitly assigned to other team members. These coordinators become the formal champions of the new safety process and are responsible for working with the steering committee to ensure the integrity of the process.

In addition to the behavioral safety coordinator, you will often want to identify an *employee safety coordinator* in each natural work team. This employee should serve to champion the process and be the point of contact for the steering committee in areas that do not have steering committee representation.

14.3 STEERING COMMITTEE'S RESPONSIBILITIES

The steering committee has the primary responsibility for implementing the design team's plans (see Table 14.2).

When you implement the behavioral safety process, get the observations started as soon as possible because they are the foundation of the process. Your steering committee can then take the time to develop a consensus and get input from area employees on the vision and values created during the design team's planning efforts. As during the design phase, you may choose to place these activities in a different order. You will often work on activities in different phases at the same time. Your steering committee, for example, may be reviewing safety data and

TABLE 14.2. Responsibilities of Steering Committee

Phase Number	Activity
1	Begin safety observations.
2	Ensure employees are trained to conduct observations.
3	Use observation data to drive improvement.
4	Develop plans for individual recognition and group celebrations.
5	Administer and deliver safety awards.

identifying safety practices that need additional effort at the same time it is working on a written statement of the area's vision and values.

Begin safety observations. The following table summarizes the steps involved in starting observations:

Step Number	Task
1	Revise safety observation checklist to meet area needs.
2	Plan kickoff meeting.
3	Conduct kickoff meeting.
4	Begin conducting observations and providing feedback.

If steering committees were not part of the design process, your implementation process should be flexible enough to allow them to modify the observation checklist to suit the needs of their work areas. On the other hand, the steering committees should be encouraged not to delete safe practices that were identified on the basis of past incidents.

Once the observation checklist is finalized, the steering committees should plan and conduct the area kickoff meetings. Generally their agenda will look much like the one recommended by the design team (see Fig. 12.1). Each steering committee will need to decide which team members will take responsibility for each agenda item. In addition, the teams will often modify the agenda to meet the needs of their work areas. Encourage the steering committees to schedule adequate time for the kickoff meetings. While it is possible to introduce the process in an hour, an ideal agenda probably requires 2 hours to allow sufficient time for employees to ask questions and discuss the process.

Here are some additional agenda items that the steering committees might want to include in their kickoff meetings:

- Role play an observation complete with feedback discussion of the safety checklist.
- Use a video of work in progress and allow employees to complete the checklist based on what they observe.
- Include participation by management and design team members.

Make every effort to ensure that all employees have the opportunity to participate in the kickoff meetings. Employees who only hear about the process second hand are more likely to misunderstand it and form a negative reaction to your efforts. Give special consideration to communicating with employees on back shifts and those who may be on vacation or special assignments.

Ensure employees are trained to conduct observations. A primary goal of the steering committee is to ensure that all employees complete the observer training workshop as quickly as possible. As discussed in Chapter 12, the logistics involved in scheduling this training for all employees may require that the workshops be spread over one or two years. Often steering committee members are responsible for conducting this training. Certainly the steering committee needs to track this training and ensure that an adequate number of trained observers are available in all areas of the facility. The steering committee will have to work with management and supervision of each area to ensure that employees are available for training. The steering committee may wish to graph the percentage of employees who have completed the training as a tool to assist in setting improvement goals and getting management support for this task.

Use observation data to drive improvement. The following table summarizes the typical steps involved for using the data to drive continuous improvement of safety:

Step Number	Task
1	Post graphs of safety data.
2	Begin reviewing observation data in safety meetings.
3	Set improvement goals.
4	Develop and implement action plans.

(Review the design team's responsibilities for developing guidelines for each of these steps in Chapter 10.)

If consistent with the design team's guidelines on the use of graphs, the steering committees should post graphs showing the observation data in work areas and locations where employees are likely to see them. A good practice is to establish a bulletin board for safety in each area. You can then readily display observation forms, safety graphs, and other safety-related information.

The steering committees should try to find a location where employees will see the bulletin board. Best are work areas, the cafeteria, the break room, and other locations with high traffic. Try to place a single bulletin board in each area. If possible avoid multiple bulletin boards that present the same information because of the time and effort required for observers to update graphs and reports on bulletin boards in different locations.

Develop plans for individual recognition and group celebrations. While the design team will usually be able to provide guidelines for individual recognition and group celebrations, the steering committee will need to plan these activities and prepare to initiate them once employees in the area begin meeting the criteria. As with the other elements of the behavioral process, the design team's plan should serve as a model for the steering committees. The steering committees may take the model process and use it as is or they may modify it extensively for the employees in their work areas.

Recognition is too important to be simply dropped or forgotten. Because many employees have had such bad experience with poorly designed award programs, some participants may try to eliminate this element of the plan. The design team should establish and communicate firm boundaries that support both individual recognition and team celebrations of success while allowing steering committees to develop plans that meet the needs of their work areas. For example, the steering committees should review the criteria established by the design team and tailor them to the area's needs. They should then review the recognition, awards, and celebration plans and decide how to make these elements meaningful to employees in each area. The ideal recognition process should provide recognition of individual contributions as part of an overall celebration of the area's achievements.

In some cases, responsibility for recognition will remain centralized with site management or the site's safety and health committee. In other cases, a separate safety awards committee may take responsibility for recognition and celebrations.

Administer and deliver safety awards. A successful award program requires that the awards be both significant and delivered in a meaningful way. The awards menu should serve as a guide, but it is based on the group's responses and may or may not reflect the preferences of a given individual. The steering committee should work with the management team to ensure that the menu remains only a guide and that each employee who meets the criteria receives an award that is meaningful and significant.

In the same way, the criteria should serve as guides. The steering committee can thus provide deserved recognition to employees who have made safety contributions not yet included on the existing list of criteria. The existing criteria should serve as a standard of comparison for considering the value of a given contribution.

The delivery is often as important as the award itself. Ensure that the person delivering the award is specific about what the employee (or team) did to earn the recognition and expresses sincere appreciation for the contribution of the employee (or team).

Delivery is important for tokens as well. If an area achieves 100 percent safety compliance and area employees earn Thank-you cards (as described Chapter 23), the supervisor or observer who distributes the tokens should personally hand each employee in the area a token, explaining why they are receiving the token and expressing appreciation. All supervisors and observers should use their imaginations to create novel ways of adding "hoopla" to such events and make the whole process fun for all involved. Handshaking, back slapping, picture taking, and providing food and beverage are all appropriate additions to a standard process.

Pay special attention to the design team's suggestions on delivering recognition. An effective delivery will make recognition successful; a poor delivery will make the recognition seem insincere and will therefore be ineffective. Again, ensure that the person delivering the award expresses sincere appreciation for the employee's (or team's) contribution and is specific about what the employee (or team) did to earn recognition. The more personal your recognition and celebrations are, the more meaningful they will be. Such items of appreciation as mugs, baseball caps, belt buckles, jackets, and the like can likewise be effective if they are handed out to individual employees with a handshake and appreciative comment. If you simply place a box of such items in the area and tell employees to help themselves, your recognition efforts will not be meaningful to many employees.

14.4 MANAGEMENT'S RESPONSIBILITIES

Management has several responsibilities for supporting the implementation efforts of the steering committee:

1. Discuss its vision of safety within the organization.
2. Discuss its expectations for implementing the behavioral safety process.
3. Participate in training on the observation process.
4. Model observation and feedback practices.

Discuss its vision of safety within the organization. Management should discuss its vision of safety with employees within the organization. Managers may discuss a formal vision or mission statement if they have developed one or they may simply discuss their personal commitment to safety and the priority of safety compared to other areas of performance. The goal is to communicate the personal commitment of individual members of management and to create a better understanding of the formal safety vision or mission statements among all employees.

Discuss its expectations for implementing the behavioral safety process. In addition to discussing its vision, management should clearly communicate its expectation that all managers and supervisors will ensure the successful initiation of the behavioral safety process. It should also clearly communicate when it expects the process to be initiated, the personal involvement expected of line management, and any other expectations regarding the behavioral safety process or how it will be initiated within the organization. There should be two-way discussions in which the managers and employees ask questions. The discussions should include a review of the implementation plans, the progress of training area management, and participation in steering committee meetings.

Participate in training on the observation process. Management should participate in behavioral safety process training. Such participation will demonstrate management's commitment to the process and help ensure that all managers also acquire the skills to conduct observations and support the process.

Model observation and feedback practices. Finally, if the design team has created a process that involves managers in the observation process, upper management should serve as model participants in the behavioral safety process. As they visit work sites within their organization, they should conduct safety observations along with managers from the work area. As part of the observations, they should provide feedback on their observations in the same way as local-level observers. Management should take special care to emphasize safe practices observed along with observations of practices and conditions that cause concern.

This topic will be discussed in much more detail in Chapter 17 which addresses safety leadership.

15 Maintaining the Behavioral Safety Process

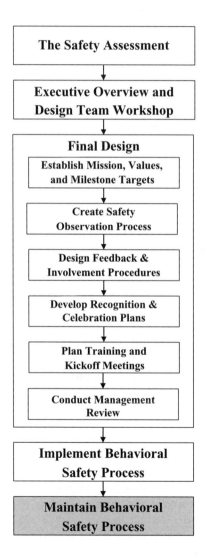

15.1 STEERING COMMITTEE MEMBERS' RESPONSIBILITIES

The primary objective of the steering committee is to ensure the integrity of the behavioral safety process. Because the primary component of this process is conducting observations, the steering committee should pay special attention to ensuring that the observation process is functioning properly. The employee safety coordinator and the steering committee must work continuously to ensure an adequate rate of observations. Steering committee members are the informal leaders in support of the behavioral safety process, and thus Chapter 17, on safety leadership, applies to them as well as to management and supervision.

Daily Responsibilities. Ideally, all steering committee members will champion the behavioral safety process. They should model practices that are aligned with the stated values and supportive of the process, including the following:

- Listen to and understand the concerns of others regarding the safety process.
- Solicit suggestions for improving the process.
- Answer questions about the behavioral safety process.
- Provide feedback to other team members and employees on:
    (a) Conducting observations
    (b) Safety practices
    (c) Alignment of behavior with values

In their daily interactions with other employees, steering committee members must listen attentively to the concerns expressed by employees, carefully noting both the content and emotions expressed in such discussions. Employees may express anger, discomfort, distrust, or other negative emotions that stem from distrust and other bad experiences. Such responses are especially likely from employees who were unable to participate in the kickoff meetings. All that you and other team members can do at this point is to ensure that you accurately understand their concerns, ensure that they fully understand the process, and solicit suggestions for addressing their concerns. Often, many of their concerns will disappear as they gain experience with the behavioral safety process.

You and your steering committee members must also be models for the kind of behavior expected of all employees, as in providing feedback to others on both daily work practices and participation in the safety observation process. Such modeling is particularly important because you want employees to understand that they should be using their feedback and observation skills all the time, not just during scheduled safety observations. Thus, whenever employees see an unsafe work practice or condition, they should try to correct the situation in the same way they would if they observed that situation during a formal observation. Getting employees to consistently interact in this manner requires that both the area management and safety team members consistently model such feedback and observation practices and provide feedback to employees on those practices.

In addition to such day-to-day interactions, the steering committee needs to pay special attention to the observation process, including reminding people to conduct their observations and, in particular, providing appreciative feedback to employees who have completed their observations. Strive to make such feedback a normal part of informal daily interactions. A simple "Thanks for completing your observations this week" will let employees know you noticed their efforts.

Weekly Responsibilities. The steering committee has two basic weekly responsibilities: the observation process and the weekly safety meetings. To fulfill these responsibilities, the following tasks must be completed:

- Developing a schedule for observers
- Conducting observations
- Planning discussions for safety meetings:
  (a) Reviewing past week's observation sheets
  (b) Identifying the week's strengths and areas of concern
  (c) Calculating and graphing the percentage of target safe behaviors
  (d) Graphing the number of observations and percentage of employees participating in observations
- Conducting the weekly safety meetings
  (a) Discussing the previous week's observation data (strengths and concerns)
  (b) Recognizing individual contributions
  (c) Creating celebrations of the area's successes
  (d) Establishing improvement targets with employees

After all employees are trained as observers, you can develop a schedule for observers for a month or even a quarter with a simple rotation of all trained observers. During observer training, until you have a significant number of employees volunteering to do observations, you will probably want observers to do observations every week.

During training the steering committee will need to complete the following tasks:

- Plan to encourage newly trained employees to conduct observations.
- Schedule new observers; then:
  (a) Explain how to complete observation forms and deliver feedback.
  (b) Conduct a joint observation, showing the observation and feedback process.
  (c) Conduct a joint observation, observing and coaching the employee's observations and feedback.

Steering Committee Meetings. Try to conduct a steering committee meeting once each month. During these meetings, the steering committee should formally

1. Are we completing the observations as we planned?
2. Have we consistently emphasized the positive (i.e., safe practices) in our discussions with employees?
3. When discussing unsafe acts and conditions, have we done so in a way that communicates our personal concern for their individual safety?
4. When discussing unsafe acts and conditions, have we done so in a way that does not embarrass or humiliate any individual?
5. Are our data from the observations meaningful?
6. Are we using the data effectively in safety meetings?
7. Are we recognizing individuals for their contributions?
8. Are we celebrating our successes as a team?
9. Have we provided observers with enough time to conduct their observations?

Figure 15.1. Sample questions to consider when evaluating a value-based behavioral safety process.

evaluate how well the behavioral safety process is working. In evaluating the process, review the percentage of observations completed on schedule. The team should also assess the accuracy of reporting, how often all employee are observed, the effectiveness of recognition efforts, and how well the feedback practices are functioning.

Your team also should provide an ongoing evaluation of the quality and integrity of its safety improvement efforts. Figure 15.1 presents a sample of self-evaluation questions for this purpose. The evaluation process might be fairly informal, based on the team members' opinions, or it might be a formal assessment conducted on a regular basis, such as annually. Formal evaluations might include interviews or written surveys with employees throughout the organization. In addition, the safety team should evaluate the interactions of its own members during team meetings to ensure that the team process remains one of openness, shared participation, and trust.

Summary of Steering Committee Responsibilities. Figure 15.2 summarizes the ongoing responsibilities of the steering committee. The employee safety coordinator (or other designated process owner) is responsible for ensuring that the team is actively addressing these responsibilities.

15.2 COMMON SITUATIONS

Figure 15.3 presents five typical situations that steering committees may discover as they assess the behavioral safety process and possible responses to these situations.

15.3 STEERING COMMITTEE'S RESPONSIBILITIES

After completing the design and implementation, the design team may evolve into an ongoing team of employees with responsibility for ensuring the maintenance and

When	Responsibility

Daily	a) Listen and ensure the understanding of concerns of others regarding safety, then address concerns constructively
	b) Answer questions about the behavioral safety process
	c) Provide feedback to other team members and other employees on:
	   • Conducting observations
	   • Safety practices
	   • Alignment of behavior with values

Weekly	a) Schedule observers and observations
	b) Conduct observations
	c) Plan discussion for safety meetings
	   • Review past week's observation sheets
	   • Identify week's strengths and areas of concern
	   • Calculate percent safe to determine trend
	   • Graph number of observations or percent safe
	   • Plan discussion

At weekly safety meetings	a) Conduct discussion of previous week's observation data b) Recognize individual contributions c) Create celebrations of area s successes d) Establish improvement targets with employees
During steering committee meetings	a) Evaluate and problem solve observations and other elements of behavioral safety process b) Assess alignment of the steering committee meeting with the team s values

When planned by steering committee	a) Plan how you will involve other employees in observation process
	b) Schedule new observers, then:
	   • Explain how to complete observation forms and deliver feedback
	   • Conduct joint observation, showing observation and feedback process
	   • Conduct joint observation; observe and coach employee's observations and feedback

Figure 15.2. Steering committee's ongoing responsibilities.

continued refinement of the behavioral safety process. The team may often assume a new name, such as *steering committee*, to reflect its new roll of maintaining the process in conjunction with the management team (see Fig. 4.1). The steering committee may have different membership from the design team. Rotating new employees onto the steering committee broadens employee participation in the process. The steering committee participants are usually employee safety representatives, often the behavioral safety process "owners" from within different areas of the site or organization. The steering committee has four primary responsibilities that may be assigned to subcommittees:

- Analyze and problem solve safety data.
- Communicate relevant safety information.

If:	Then the steering committee should:
Number or percent of observations is below target	1. Ensure that supervisors and observers are trained in the observation process and what s expected 2. Enhance feedback and recognition for completing observations 3. Ensure that supervisors are evaluated on observations completed in their work areas
Observation data suggest safety is decreasing (downward trend in percent safe)	1. Ensure effectiveness of problem-solving discussion in area safety meeting 2. Ensure discussion results in commitment to improve specific safety practice or practices 3. Consider changes in observation process to increase effectiveness of observations (i.e., more observations, more employees involved, and other steps)
Observation data suggest safety is increasing (upward trend in percent safe)	1. Arrange recognition or celebration of improvement in area safety meeting
A few employees are negative about the behavioral safety process	1. Ensure that the negative employees have participated in orientation and training 2. Involve those employees in observations and in steering committee
Most employees are negative about the safety process	1. Ensure that observers are stressing positive factors when providing feedback on observations 2. Ensure accuracy of observation data 3. Increase use of safety awards and recognition based on observation process

Figure 15.3. Common problems found during evaluation and suggested solutions.

- Address employee concerns.
- Coordinate recognition.

Analyze and problem solve safety data. The steering committee analyzes safety data to find meaningful information. The analysis should include identification of trends and Pareto analysis of both observation data and data on injury rates. The data may identify areas where employees are not maintaining the observations or suggest safety practices that need additional training or attention. Often the steering committee will publish companywide data comparing the data from observation categories with the causes of incidents that resulted in injury. Depending on the size of your organization, such data will frequently show a clear correlation between the practices with a low percentage of safe practices (or high percentage

Category	Safety Practice	Percent Unsafe	Percent of Incidents
Body position	Steps to avoid burns	16	15
	Proper pulling and lifting	24	11
	Avoids pinch points	8	7
	Repetitive motion	4	4
	Eyes on path	0	4
	TOTAL	52	48
Tools and equipment	De-energize	8	4
	Proper use of tool	12	4
	Other	8	7
	TOTAL	28	15

Figure 15.4. Sample statistics comparing observation data with causes of incidents.

of unsafe practices) and the causes of injury incidents. Figure 15.4 shows an example of a comparison of data on unsafe practices with causes of injury incidents. The steering committee may also work with other steering committees to ensure that the process is working effectively and to address problems such as those discussed in the previous section.

Communicate relevant safety information. The steering committee also communicates relevant safety information to employees throughout the company, often using either a newsletter or area bulletin boards. Usually this includes publishing the results of the steering committee's statistical analysis in tabular and graphical form. It may also include additional information that supports the behavioral safety process, such as discussion topics or training materials for safety meetings.

Address employee concerns. Another steering committee responsibility is evaluating and responding to employee concerns about the behavioral safety process and other safety-related issues. Often observers will note employee concerns during observations or identify safety concerns or questions about safety that go beyond the responsibility of the area. Employees may also raise such concerns during safety meetings. The employee safety coordinator should ensure that such concerns are passed along to the steering committee for a response.

Coordinate recognition. The steering committee may also be responsible for planning and coordinating individual recognition and area celebrations. The steering committees may, for example, nominate individuals for recognition on the basis of a particular contribution to area safety. The steering committee may evaluate such nominations for sitewide recognition, determine whether such recognition is justified, then plan recognition appropriate to the contribution. This role helps ensure that recognition is consistent across the organization.

15.4 MANAGEMENT'S RESPONSIBILITIES

Maintaining the observation process and other elements of the behavioral safety process requires an ongoing effort from both management and steering committees.

142 MAINTAINING THE BEHAVIORAL SAFETY PROCESS

When	Responsibility
During visits to work areas	a) Conduct observations and provide feedback using area s observation sheet (model emphasis on positive feedback) b) Review and provide feedback on completion of observations c) Discuss process with steering committee members d) Provide appreciative feedback to: • Steering committee members who champion the process • Employees who complete observations • Area management and all employees for improvement, goal attainment, and other area successes
At management meetings	a) Review and provide feedback on percent of completed observations in each area (not the percent safe resulting from observations) b) Discuss plans for further process improvement c) Identify individual contributions worthy of recognition
During performance reviews with managers and supervisors	a) Assess level of support and participation in behavioral safety process (percent of completed observations, not the percent safe resulting from observations)

Figure 15.5. Management's responsibilities for maintaining a value-based behavioral safety process.

Implementation is not like walking into a room and flipping on a light switch. Keeping the process alive requires sustained attention to the observations, use of the data, and recognition. Figure 15.5 provides guidelines on management's responsibilities to ensure effective maintenance of the behavioral safety process. This topic is explored in more detail in Chapter 17.

16 Some Final Suggestions on Implementation

Here are seven suggestions to improve your chances of having a successful behavioral safety process.

Suggestion 1: Don't use a cookbook approach. Although this book provides some guidelines, your success depends on your ability to apply these guidelines in the context of your organization's culture. This suggestion pertains not only to the approach described in this book but also to comparable approaches from consulting organizations around the country. Do not simply buy a package. Make sure you use the key elements of the approach to develop a system that meets the needs of your organization. In addition, do not be too quick to discard key elements just because they are difficult to implement or maintain in your organization.

Suggestion 2: Plan and clearly define management's role. The observation process requires time and a great deal of effort. It will require active support from all levels of management. Pay special attention to suggestions for involving management and defining management's role in the process. Management's role will be the critical factor in both the long-term success of your observation process and the day-to-day elements of your safety improvement process.

Suggestion 3: Maximize participation in the final design. The only way to create ownership is through meaningful involvement in the design process. This process requires a high level of participation and provides several options for ways of involving people. Do not make the mistake of designing a safety process in a vacuum, then trying to implement the program by mandate. Involve people in the design at each stage. Then have those who assist with the design take it back to their work areas and get input and suggestions from their colleagues.

Suggestion 4: Create a different checklist for each area. The research studies used checklists of specific safe behaviors that were job and area specific. Unless you are in a small facility, do not try to develop a generic checklist that works for all work areas. Maintenance has different safety requirements than a laboratory, for example, and their respective checklists need to be different. To maximize the value of the checklists, they should be explicit enough to address the specific safe practices of different job functions.

Suggestion 5: Don't create a bureaucracy around the data. The value of this process is in getting everyone to pay attention to on-the-job safety, and creating a paper storm will hinder any process. Build informal systems of accountability based on the observable parts of the system. Do not create an elaborate system of paper reports. Do pay attention to the safety process during informal contact with individuals in the work areas and during formal meetings at each level of the organization.

Suggestion 6: Use classroom training only when needed. Place emphasis on designing a training process that satisfies your needs, not putting all employees through extensive classroom training. Provide enough training to create the understanding that people need to support the process. Also, do not think of training as strictly a classroom process. When training observers, for example, giving them an opportunity to actually conduct an observation is often more effective than classroom training alone. Use training only when appropriate, and select an appropriate process for delivering the training that is needed.

Suggestion 7: Persevere. Don't quit—ever! False starts typically characterize the implementation of any significant new process. Implementation is often two steps forward and one step back. The key to success is continuous improvement. Learn from each step so that you can do it better the next time. Just keep fine tuning your process until you achieve zero incidents. Then strive to maintain that level of safety excellence.

17 Special Topics: Safety Leadership

Almost everyone agrees that management support is the single most important element for ensuring a successful behavioral safety process. The exact nature of that support, however, is the topic of some debate. Much of the debate stems from the fact that different organizations need management to participate in different ways. The role that management and supervision play in supporting behavioral safety should ultimately be tailored specifically to the unique needs of each organization and its safety process. As with the other elements of behavioral safety, one size simply does not fit all.

Regardless of their role in behavioral safety, managers and supervisors are responsible both for monitoring and enforcing safety rules and procedures and for ensuring that the workplaces under their charge are properly designed and maintained. Behavioral safety is much easier in organizations where employees see their leaders consistently promoting and paying attention to safety.

17.1 BIGGEST BARRIER TO EFFECTIVE SAFETY LEADERSHIP

The biggest barrier to effective safety leadership in many organizations is found in its accountability and measurement systems. Typically, the management operations within organizations include a measurement system that overemphasizes production, cost, and even quality over safety. Many organizations track production on an hour-by-hour basis, and managers and supervisors watch these figures carefully across the workday. As discussed in the chapter on culture, these systems are out of balance in that they create an unbalanced emphasis on production. The measure of safety is often whether or not the organization has had an incident. If incidents are rare events, production and other more frequent measures get most of management's attention. In paying attention to these measures, management—often unintentionally—sends the message that production is what is important to the organization while safety gets little attention, sending the further message that it is not important. The behavior of managers and supervisors is a function of the system in which they work. The adage "What gets measured, gets done" applies as much to them as to the employees within the system. In such a system, managers spend far more time paying attention to production and too little time paying attention to safety. This state of affairs contributes to several other barriers to effective safety leadership.

17.2 OTHER BARRIERS TO EFFECTIVE SAFETY LEADERSHIP

In our experience, four related barriers frequently prevent managers from being more effective in promoting safety:

- Management by exception
- Inadequate monitoring and follow-up
- Management by fear
- Overemphasis on statistics

Managers and supervisors often rationalize that managing by exception is the most cost-effective use of their time. They seem to believe they need to pay attention to safety only when someone gets injured or when they identify an unsafe practice or condition. This approach automatically places the organization in a reactive mode. Managers with this philosophy respond to correct a problem only after a problem occurs, otherwise spending their time on what they see as more urgent issues. Many of these same managers would never dream of managing quality or production in the same way. Imagine a modern manager who responds to quality only when defects appear! Such a system virtually ensures that defects will continue to occur. In such *management-by-exception* systems, safety gets very little ongoing attention. Managers and supervisors are attending to more urgent brush-fires, at least until an injury occurs. The real problem with this approach is that it will not achieve the level of safety most exemplary organizations strive for. In exemplary organizations, "one strike and you're out." Thus, once an injury has occurred, the chance of achieving a safety target is either very difficult or even impossible.

A second and closely related barrier to effective safety leadership is simply not spending sufficient time monitoring safety practices and systems. As with management by exception, this barrier results from such factors as inadequate management/leadership training, poor self-management, or too many competing activities (paperwork, email, telephone calls, meetings, etc.), but the basic problem is that managers and supervisors do not give adequate time to monitoring safety practices at each level of the organization. They may be spending too much time in their office and not enough time directly observing and discussing work practices. Or they may attend to and discuss other aspects of the work without noticing or discussing safety practices or conditions. This problem can also be related to reports and meeting minutes related to safety: If a manager does not regularly review, discuss, and act on information from both direct observation and safety reports, employees begin to believe that safety is not important.

Another common barrier relates to managers who rely on the threat of discipline to motivate employees to work safely. The illogic of this approach is almost self-evident—if employees are not going to work safely to avoid injury, why would they work safely to avoid disciplinary action? The situation is about the same as relying on the fear of injury to motivate employees to work safely. The probability of disciplinary action is frequently too low to motivate a consistent level of safe work practices. Furthermore, in this type of work environment, employees become very good at identifying times (such as when certain managers are present) when discipline is likely and times (such as the evening shift) when discipline is less likely. (The problems with overreliance on punishment are discussed in much more detail in Chapter 1.)

The final barrier to effective leadership is an overemphasis on incident rates. Again, this is in essence a variation on management by exception, as it places leadership in a reactive mode. The problem is that emphasizing safety in terms of the rate of injuries creates a perception that the company is more concerned with statistics than with the safety and well-being of the employees. Management surely needs to be concerned about results and incident rates related to safety, but it has to balance this concern with attention to safety on the job. (The problem of overemphasizing results is discussed in more detail in Chapter 4.)

17.3 LEADERSHIP'S SPECIAL ROLE

Ideally, before implementing a behavioral process, managers and supervisors send a clear message to employees about the importance of safety by frequently paying attention to safety and making decisions that clearly show their commitment to safety. Part of this responsibility is to ensure that all elements of a traditional safety management system are in place. These include regularly attending to safety procedures, conducting safety audits, modeling safety in their personal actions, and ensuring compliance with all regulatory requirements.

Leaders communicate what they value through what they say and do. This is often referred to as "say–do" correspondence. The matrix presented in Table 17.1 graphically represents this correspondence. While important for all employees, this correspondence is even more critical for those in leadership roles. A manager in an organization can talk about the importance of safety or not, and the manager can act in a safe manner or not. Table 17.2 provides examples of the kinds of comments and actions that show a manager values safety in an organization.

TABLE 17.1. Say–Do Matrix Showing Model for Assessing How Supervisors Demonstrate Support for Safety in What They Say and Do

		Act safely	
		Yes	No
Talks safety	Yes	Recognize Reward Promote	Train and coach
	No	Train and coach	(Preselect) Train Coach Terminate

TABLE 17.2. Examples of What Leaders Say and Do in Support of Safety

Examples of What Leaders Say	Examples of What Leaders Do
Always start meetings with safety items.	Ensure safety work orders get addressed.
Ask for direct reports about what has been done or planned to improve safety.	Provide recognition to those who champion safety.
Ask how the behavioral safety process is going.	Arrange and participate in celebrations for safety achievements.
Ask about the behavioral safety process, including	Model safe practices in personal actions.
• Level of participation	Conduct behavioral safety observations.
• Number of observations	Monitor safety processes.
• Steering committee's action plans	Ensure incident investigations.
Provide feedback on safety practices.	Ensure employees have time to participate in safety improvement efforts.
Include safety in job assignments.	Conduct regular facility audits to identify and address conditions.
Balance questions on safety actions and results.	Review behavioral safety process data.

If an organization values safety, it will respond as indicated in the cells represented in Table 17.1. Effective leaders who both verbalize the importance of safety and act in ways that support safety need to be recognized, rewarded, and ultimately promoted (assuming that they are performing well in other areas). However, if a leader is not talking with employees about safety, promoting safety practices, providing feedback, discussing the importance of safety in safety meetings, and verbally supporting safety in other ways, an organization that values safety will ensure that the leader gets training and coaching that encourage more verbal support of safety. Similarly, if a leader is not acting in ways that demonstrate the importance of safety through personal actions such as paying attention to safe and unsafe work practices, consistently following safety procedures, following up on safety-related work orders, ensuring that employees have time to participate in safety meetings, and the like, again, the organization that values safety will ensure that additional training and coaching are provided to help the leader "show the way" through personal actions. Hopefully, potential employees who did not express support for safety or otherwise indicate its importance would have been screened out during the hiring or selection process. If a leader were to fall into this cell of the matrix, an organization with an effective safety culture would address the leader promptly through training and coaching or, in extreme cases, termination.

17.4 PHASES OF MANAGEMENT SUPPORT

The role of management in support of behavioral safety changes over time. The first step for management is to decide to implement behavioral safety and commit the resources to support the planning and implementation. A common mistake at this stage is underestimating the resource requirements. Management needs to fully understand both the time and money required to implement behavioral safety effectively. It often mistakenly believes implementing behavioral safety simply involves training employees to conduct safety observations while failing to understand the planning and training required to create a system that will support the observation process for long-term success. At this stage, management's role is to learn more about behavioral safety and to understand the importance of creating a process of employee involvement that is formally supported by the organization.

Once the decision is made to provide the necessary support, management's role is to monitor progress through the remaining stages of planning, implementation, and maintenance of the process. While the monitoring role remains the same through each of these three stages, attention needs to be focused on different elements in each stage.

17.5 MANAGEMENT'S MOST IMPORTANT ROLE

Managers who want to support a behavioral safety process need to pay attention to their process. Leadership research has demonstrated that the single most important activity that distinguishes effective from mediocre leaders is the amount of time they spend in monitoring the performance of their organizations (Komaki, 1998). The implications of this research are clear. Managers who want to support a behavioral safety process should monitor their process at each stage of its development: planning, implementation, and maintenance. They need to stay in touch with the progress of the process during planning and implementation, then how well the process is functioning once it has reached the maintenance phase. Generally, this means they should review plans for the behavioral process before implementation, then get regular updates once the process is in place.

Managers can monitor their behavioral safety process both formally and informally. Informal conversations may be the more critical of the two. Peters and Waterman (1982) suggested that, to be effective leaders, managers needed to practice "management by walking around," meaning they needed to get out of their office and tour their operations. Komacki's (1998) findings support this recommendation while further suggesting managers need to do more than merely walk around. They also need to talk with employees about their performance.

These studies suggests that the way managers and supervisors can best pay attention to behavioral safety is to regularly ask questions about the process in their areas of responsibility. This might be described as the "tell me about it approach" to initiating conversations about how well the process is functioning. These discussions should be conversational, usually starting with broad, general questions and then moving to more specific details.

While almost everyone agrees that safety is important, it is often not urgent. For example, a crane inspection can seemingly always be done tomorrow, while getting the production line back up and running has to be done right now. One of our challenges in safety is to create a system that establishes deadlines that helps establish the urgency of management practices that promote safety.

A contribution often attributed to Du Pont is the importance of placing safety first on the agenda of every meeting. A decade ago, this was an important step in communicating the importance of safety to employees at all levels of the organization. Unfortunately, for many organizations, this practice has been translated into a single question, "Did we have any incidents?" If the answer is no, the meeting moves on to the other items on the agenda. If this is how your management is currently discussing safety, your organization is probably about average, a "C student" in safety at best. While it increases attention to safety, this approach still demonstrates a reactive, management-by-exception kind of approach.

In exemplary organizations, the management team also discusses what is being done to address safety. Additional questions might be something like "What have you done this past week to promote or improve safety?" and "What are you planning to do this coming week about safety?"

17.6 POSITIVE QUESTIONS

The ability to ask good questions about performance appears to be an important skill for leaders wishing to demonstrate their interest in supporting behavioral safety. We have also seen leaders who ask ineffective questions in their efforts to monitor performance. The difference is that effective questions make employees feel that the leader cares about what is happening while ineffective questions make them feel interrogated or grilled. Positive questions are helpful while ineffective questions make listeners feel defensive. Positive questions are likely to prompt openness and honesty while ineffective questions are likely to prompt listeners to provide excuses, rationalizations, and as little real information as possible.

In trying to find the difference between these two types of questions, we have noticed that managers who ask questions that start with "what" or "how" get better responses than managers that ask questions that start with "why." We have also found this to pertain in training observers in behavioral safety. While we are always very interested in having observers identify the root causes of at-risk behaviors, asking "why" nevertheless often puts employees on the defensive.

In addition, second-person ("you"-oriented) questions are more likely to get listeners to talk about their performance. Thus, a monitoring question early in the discussion for someone who is struggling with a safety problem might be something like "What have you tried so far?"

Listeners also respond positively to questions that seek solutions and focus on the desired outcome or behavior. Such questions are more likely to be perceived as helpful while questions that are focused on the problem are perceived as seeking to find fault or place blame. Positive questions often quickly shift the focus from what has been done in the past to the future, such as current plans or what needs to be done. An effective question in continuing the discussion following the response to the question in the previous paragraph might be "What are you planning to do next?"

In creating a natural dialogue, the exemplary leader asks questions that fit the situation. Effective questions come naturally in the context of direct observations.

TABLE 17.3. Questions Leaders Might Ask to Show Interest in Behavioral Safety Process

Questions For:	Might Include:
Supervisors and managers, steering committee members	How is the process working? What is working well? What are the challenges? What percentage of employees are participating in conducting observations? How many observations have been done in the past week or month? What is working well? Who has been active in supporting the process? What can the leadership do to help?
Employees who conduct observations	What kinds of safety practices are people performing well? What kind of safety concerns are they seeing? How are associates responding to the observations?
Other employees	Are employees planning to participate in the process? What do they like about the process? What are the barriers that prevent them from conducting observations?

Accordingly, leaders can easily ask effective questions that fit the situation in the context of touring work areas. Positive questions and resulting feedback then flow naturally and build on one another as the conversation progresses. Leaders who use questions effectively also often start by seeking broad general information, then gradually drill down into more specific details.

Used correctly, effective questions build a better understanding for everyone involved. The manager gains a more accurate understanding of how well the process is functioning, and at the same time employees gain a better understanding of what is important to the manager. Ideally, by regularly creating this type of dialogue between management and employees, both gain additional insights that can contribute to continuous improvement, which is the purpose of effective leadership in the first place. Table 17.3 provides some questions that leaders might use in interacting with others to learn how well their behavioral process is functioning.

17.7 FORMAL MONITORING AND MANAGEMENT ACTION ITEMS

Management may also wish to stay informed through more formal reviews of the process. Such reviews may include examining summaries of observation data or getting updates from a steering committee representative during staff meetings. The updates should include the current priorities established by the steering committee and a description of what the committee is planning to do to address those priorities.

The update should also include requests for management action needed to support the steering committee and the behavioral process. The steering committee often needs management assistance to address facilities issues contributing to unsafe actions that the steering committee is targeting for improvement. It is also likely to need assistance in arranging celebrations and recognition events. For example, if the steering committee is focused on improving lifting practices, it might ask management to consider the matter of a lifting device that is difficult to use, causing employees not to use it when they must lift something quickly. The idea is not to create a laundry list of items for management to address, but to ask for assistance with specific conditions and facility issues related to the steering committee's top priorities or goals.

Management also needs to consider other actions needed to support the process. One of its key activities is providing recognition and expressing appreciation for actions taken by individual steering committee members. Among this committee's functions is arranging recognition and celebrations for other employees to help sustain the safety process, whereas management provides recognition to the committee's members for their efforts. Sending steering committee members to professional conferences, such as the annual Behavioral Safety NOW Conference, for example, is one way to recognize individuals who have championed the process.

17.8 ROLE OF MANAGERS AND SUPERVISORS IN OBSERVATIONS

One of the more controversial questions in behavioral safety has been whether or not managers and supervisors should participate in conducting observations. Even today, many practitioners recommend that observations in a behavioral safety process be strictly limited to employees. As discussed in Chapter 9, the decision as to who should conduct observations has to be determined by the needs of a particular organization. Except in organizations with a high degree of conflict and distrust between employees and management, managers and supervisors should participate in conducting observations. This approach clearly sends the message that everyone who comes in the gate has a shared responsibility for safety.

Our research shows a high correlation between management and employee participation (Cook and McSween, 2000). When managers and supervisors participate, a larger percentage of employees participate. Companies that include everyone in the process of conducting safety observations experience improved communication between levels of employees and also develop a better understanding by managers and supervisors of safety issues faced by employees. This understanding results in better decisions regarding resources in support of safety.

One example is that of a supervisor participating in the behavioral process in a gas production organization. During one of his first field observations, this supervisor is surprised to see his crew using a short pipe wrench with a cheater bar. He intervenes to prevent one of the crew from standing on the cheater bar. Without the threat of disciplinary action during behavioral observations, the crew feels safe in showing the supervisor how the job is typically done in the field. The supervisor recalls that he had previously denied the purchase of new wrenches with longer handles and subsequently approves the purchase of more appropriate tools. As illustrated by this example, the observation process allows supervisors who may be under pressure to cut costs to be better informed about actual practices in the workplace.

In addition to developing a higher level of trust between employees and managers, involving managers and supervisors in behavioral observations requires that training for observers be very precise in order to avoid any potential misunderstanding. Generally, management assures employees that behavioral observations will not be used as the basis for disciplinary action. Potential conflict nevertheless occurs because supervisors and managers have a legal obligation to ensure compliance with company rules and procedures. The best resolution may be for supervisors and managers to ensure that employees are not engaging in activities that would warrant disciplinary action prior to beginning an observation. This is almost always a matter of extensive discussion by the design team, but in practice, disciplinary action is such a rare event that this is a nonissue for most organizations.

When disciplinary policies and requirements are well understood and consistently applied, employees will understand that, when it does occur, discipline is a necessary enforcement of "life-critical" safety rules. They will understand, for example, that a person who is intoxicated at work probably faces termination. They will also understand that someone who fails to lock and tag out equipment is potentially putting their life in danger and that such failure calls for discipline. If, on the other hand, disciplinary policies and requirements are not well understood or consistently enforced, management needs to address these issues as a problem within the management system, independent of the behavioral safety process.

17.9 IMPORTANCE OF INFORMAL LEADERS

If possible, every employee work group should have a person designated as the behavioral safety representative. Ideally employees volunteer for the position, and one of the volunteers is selected by the co-workers. The title for this position varies in different organizations but is typically something like employee or crew safety representative. These persons are the point of contact for the behavioral safety process for a crew or on team and are responsible for promoting the process on that crew or team. They have a special obligation to conduct behavioral safety observations with the frequency planned by the design team, thus modeling participation for their co-workers. They will often have related safety duties, such as ensuring that observation forms are available in the area, reviewing behavioral safety data in safety meetings, posting or updating graphs, arranging recognition and celebrations with supervision, and other tasks in support of the process. In many respects their responsibilities resemble those of formal leadership (i.e., managers and supervisors) except that they do not have the formal authority and enforcement responsibilities of those positions.

In small organizations, these representatives are usually also steering committee members. In larger organizations, however, they constitute a separate group in order to avoid the steering committee becoming too large to function effectively. When the organization is a large one, the design team also needs to establish how the steering committee will communicate with the safety representatives. Often, each steering committee member conducts a monthly meeting with a group of safety representatives to ensure two-way communication in support of the behavioral safety process. These meetings may also be used to solicit information for the steering committee or to allow the steering committee to send information, such as priorities and action plans, to other areas. In some cases, much of this communication can be done electronically through email.

17.10 OTHER LEADERSHIP RESPONSIBILITIES

Two other safety responsibilities of management need emphasis. Management and supervisors need to be actively involved with ensuring that those who participate in the behavioral safety process are reinforced for doing so. Management shares responsibility with the steering committee to ensure that individuals who champion safety get appropriate recognition and that teams have an opportunity to celebrate their successes. All of the suggestions in Chapter 11 are as relevant to management as they are to the steering committee.

Another area that often needs management attention is employee orientation. In particular, supervisors should provide a formal orientation to employees new to their department. Too often supervisors underestimate the importance of providing clear instruction about safety practices such as proper lifting or cutting techniques in the context of a new employee's job. The ideal orientation should include a new employee orientation checklist and an introduction to the behavioral process to help ensure new employees learn about the hazards in their areas.

17.11 SAFETY LEADERSHIP CHECKLIST

In some organizations, the design team or the management team creates a self-observation checklist for managers and supervisors. This checklist typically includes a list of the behaviors that leaders need to engage in to support the behavioral safety process. Managers and supervisors may be required to complete this checklist in place of or in addition to normal behavioral safety observations. These checklists are typically turned in and summarized in the same way as other observation checklists. The summary data may be reviewed as part of the steering committee meeting or summarized and reported to the site management team. Figure 17.1 presents a sample self-observation checklist for managers and supervisors.

Safety Leadership Checklist			
Name: _____ Date: _____			
Behavior	Yes	No	n/a
1. Conduct a behavioral safety observation			
2. Review observation data in safety meeting			
3. Encourage at least one employee to do safety observation			
4. Correct an unsafe condition			
5. Conduct one informal area safety review this week			
6. Hold celebration when group meets goal			
7. Give positive feedback at least 1 a week (safety related)			
8. Provide feedback to employees on safety behaviors			
9. Complete new employee orientation			
10. Completed checklist returned to Safety by Friday			

Figure 17.1. Example of a management self-observation checklist.

17.12 CONCLUDING COMMENTS ON LEADERSHIP'S ROLE

The active and visible support of management and supervision is critical to the long-term success of a behavioral safety process. Their role is two-fold: (1) to show their personal commitment to safety through what they say and do and (2) to pay careful attention to the behavioral safety process without taking it away from the employees. This means they must monitor the process in a way that communicates their interest and support without becoming directive or using their authority to mandate changes in the process. To do so, they must ask good questions that communicate a sincere interest without making employees feel like they are being interrogated. They must also model participation and conduct observations just as the design team planned them.

Table 17.4 compares practices of effective leaders in a behavioral safety process with activities of ineffective leaders. By asking good questions, effective leaders build involvement and develop their employees while less effective leaders tend to tell their employees what to do. We have seen several instances where a manager has mandated a change in a safety team's plans even though the plans were aligned with the team's safety objectives within the boundaries established when the team was chartered. Such mandates are particularly destructive when a team of employees has been working on an issue and the manager does not take the time to understand the planning activities that lead to the team's recommendations.

In addition, leaders must support the steering committee, including assisting them in addressing facilities issues and conditions that pertain to the chief areas of safety concern. Leadership must actively support and participate with the steering committee in (1) providing recognition to employees who champion safety and

TABLE 17.4. Practices of Effective Versus Mediocre Leaders

Effective Leaders	Mediocre Leaders
Conduct safety observations in work areas	Do not regularly conduct safety observations
Monitor: • Activities in their areas • Steering committee plans • Safety observation and outcome data	Do not pay attention to their process
Lead by: • Example • Effective questions • Input and suggestions	Lead by mandate
Build for long-term involvement	Push for immediate results

(2) arranging celebrations for teams that achieve success. Finally, the leadership team must ensure that steering committee members and employee safety representatives are properly recognized for their roles in the process. These actions will ensure that the discretionary efforts of the steering committee and employee safety representatives continue, thereby ensuring that the behavioral safety process will be long-lasting.

18 Special Topics: Serious-Incident Prevention

Witnessing a catastrophic event or experiencing the consequences of such an incident can make a lifelong impression. The impact can be even more profound if a catastrophe resulting in fatalities or other serious consequences occurs in an operation for which you are responsible.

This was the case for a division manager attending a meeting of a major chemical facility's safety committee prior to his retirement. Even though the vapor cloud explosion that resulted in three fatalities in the manager's division had occurred nearly two decades earlier, its impact on the manager was still evident. In his final comments to the safety committee, the manager recognized the importance of continuing to drive down the facility's OSHA-recordable case rate. However, his primary message was a challenge to the organization not to lose sight of its critical responsibilities for preventing serious, high-consequence events that forever change lives. The incident had clearly changed this manager's perspective permanently, and he hoped others would continue to diligently apply those lessons of the past to prevent serious incidents from occurring in the future.

Too often, the lessons learned from such events are short-lived. With the passage of time, organizations find that the actions critical to maintaining safe operation are once again being ignored. Organizations need a systematic management approach to ensure that actions critical to preventing serious incidents are accurately identified and diligently executed over the long term.

Performance management is a process that incorporates (1) employee involvement, (2) measurement of upstream performance indicators, (3) performance feedback, and (4) reinforcement contingent upon performance. It has provided a framework for significant improvements in key performance areas, including product quality, productivity, customer satisfaction, and safety. As described in other chapters of this book, such processes have formed the basis for successful behavioral safety initiatives leading to breakthrough levels of improvement in injury reduction for many companies. Performance management is the foundation for the proven management approach described in this chapter. The serious-incident prevention process has proven effective for achieving and sustaining serious-incident-free operations.

Note: This chapter was written by T. E. Burns, Quality Safety Edge, and based on his book *Serious Incident Prevention*, 2nd ed., Butterworth-Heinemann, Woburn, MA, 2002.

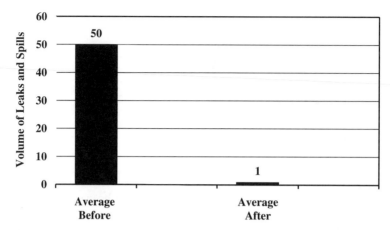

Figure 18.1. Average volume of leaks and spills at a chemical tank farm operation for three years before and after implementation of a behavioral approach.

These techniques have been successfully implemented to achieve major reductions in incidents such as serious injuries and fatalities, accidental releases of chemicals, hazardous material transportation incidents, regulatory agency violations, and security breaches. To illustrate, Figures 18.1 through 18.3 summarize the results of the improvements achieved in one division of a major petrochemical company through the implementation of the eight-element serious-incident prevention process described in the pages that follow. Each figure compares the average level of performance for the three years immediately prior to implementing the serious-incident prevention process with the breakthrough levels of performance achieved during the three years immediately following the implementation of the improved prevention process.

Success in preventing serious incidents over the long term is firmly linked to an organization's capabilities for identifying the tasks critical to success, then successfully managing the many details involved in properly executing these tasks.

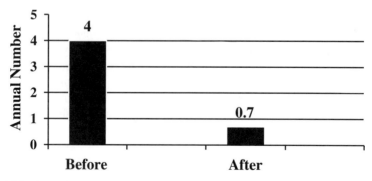

Figure 18.2. Annual number of nonaccident tank car releases for three years before and after implementation of a behavioral improvement process.

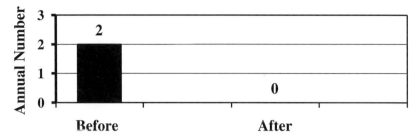

Figure 18.3. Annual number of regulatory agency violations in a pipeline operation for three years before and after implementation of a behavioral improvement process.

Success can best be achieved through application of a process model that merges proven performance management techniques with sound risk management practices. Experience indicates that the following eight process elements are essential for effectively maintaining workplace conditions and practices necessary to sustain incident-free operations.

18.1 ELEMENT 1: BUILD MANAGEMENT COMMITMENT AND LEADERSHIP

Implementing a serious-incident prevention process is generally a management-driven improvement effort. Management commitment and leadership are critical to overcoming barriers to success and for maintaining recognition of the serious-incident prevention process as a top priority throughout the organization. However, even the lack of a clear upper management mandate to implement an improved serious-incident prevention process should not be considered an insurmountable barrier for the individual manager or supervisor who identifies an opportunity for improvement.

Once a manager or management team decides to increase efforts to prevent serious incidents, they need to educate themselves on this overall approach so that they understand each of the steps outlined below. Supervisors and managers at each level of the organization need to recognize that they are considered "top management" by their subordinates and that each level of management has the capability to assume a strong leadership role. In many organizations, self-directed managers who have a strong desire to make a true difference seek out, work out, and require the procedures needed to improve workplace safety. Such perceptive managers do not require that a tragedy occur before taking needed actions. They understand that the organization must be proactive in taking the steps necessary to prevent serious incidents.

18.2 ELEMENT 2: INVOLVE EMPLOYEES

Full employee involvement is essential to leveraging the organization's limited resources, harnessing employee knowledge, and facilitating employee ownership of the safety process. Parents typically show polite interest in other people's babies, but they reserve real sacrifice and passion for their own. The development of employees' passion for workplace safety improvement requires comparable active involvement and ownership. The full benefits of employee involvement can be achieved when, instead of simply participating, employees assume leadership roles in developing, implementing, maintaining, and improving the serious-incident prevention process.

18.3 ELEMENT 3: UNDERSTAND THE RISKS

Success in any endeavor requires knowledge of potential risks. Without a thorough understanding of the risks, the need for safety concerns and prevention is unfortunately often recognized only after a serious incident has occurred.

In view of the nonroutine nature of serious incidents, the focus must be on *what can happen* rather than on *what has happened* in the past. The argument that a unit has been operating for many years without problems is often offered as rationalization for a lack of concern. However, such a misguided point of view must be granted only limited consideration in evaluating the potential for a future incident. A systematic process involving all levels of the organization must be in place for identifying the risks that can lead to incidents that bring serious consequences for the organization.

18.4 ELEMENT 4: IDENTIFY CRITICAL WORK FOR CONTROLLING THE RISKS

A key to successfully managing serious safety risks is proactive identification and execution of tasks that are critical in controlling risks. In many operations these tasks are extensive, and hence an effective system must be established to ensure that all of the proper conditions and work practices are in place and maintained for supporting the diligent execution of the work over the long term. The specific list of critical tasks required for safety success will vary depending upon each organization and the types of risks that must be addressed. Typical examples of these tasks include training, inspections, observation and feedback, equipment testing, emergency drills, and preventive maintenance. The serious-incident prevention process model includes the identification of the critical tasks that an organization must focus on to successfully control major safety risks.

18.5 ELEMENT 5: ESTABLISH PERFORMANCE STANDARDS

Once the tasks critical to incident-free operations are identified, standards of performance are required to establish expectations for satisfactorily executing them. Performance standards must be based on careful research in order to provide guidance in sufficient detail while avoiding overly excessive requirements that increase costs without a corresponding benefit in safety performance.

Questions often arise within an organization on the proper frequency for performing tasks, as in how inspections, audits, hazard reviews, and training need to be conducted. Additional questions may arise regarding the appropriate qualifications and training for employees who perform equipment testing, conduct inspections, or train others. Standards that do not require employees to perform safety tasks frequently enough or that allow unqualified employees to perform critical tasks fail to provide an adequate margin of operational safety. The serious-incident prevention process model emphasizes the need to establish standards that are effective in both preventing incidents and fully utilizing resources.

18.6 ELEMENT 6: MAINTAIN MEASUREMENT AND FEEDBACK SYSTEMS

Success in any endeavor, including serious-incident prevention, requires more than knowing what to do and when to do it. Unless the tasks critical to safety success are executed as planned, the organization's efforts will simply become another initiative with "good intentions" that failed rather than a successful one providing an outstanding long-term return on the investment of time and resources. Performance measurement systems will be required to effectively monitor performance in properly executing the actions required to implement the serious-incident prevention process. Further, effective feedback systems must be established to communicate progress on key initiatives to the employees responsible for performing the work critical to success.

Too often after an incident, managers find that critical work practices have been inappropriately altered or terminated. The effective measures of upstream performance indicators established by the serious-incident prevention process help ensure that critical actions are sustained for the long term. The establishment of effective feedback systems provides for timely adjustments to the safety process in proactive actions taken before the occurrence of an incident rather than reactive actions after damage has been done.

18.7 ELEMENT 7: REINFORCE AND IMPLEMENT CORRECTIVE ACTIONS

People tend to sustain activities when they feel positively reinforced for their efforts. Unfortunately, positive reinforcement for employees in the workplace for executing the tasks necessary for serious-incident prevention is often all too rare. In fact, completing tasks required to maintain safe operations may at times seem "punishing" to those responsible for performing them by slowing them down in their efforts to achieve results in areas where the real "glory" may lie (e.g., increased production, productivity, or cost savings). Managers who depend on employees to execute tasks in a timely and thorough manner that are critical to the prevention of serious incidents must ensure that a process is in place to positively reinforce responsible employees for taking and maintaining proper safety actions. Such positive reinforcement is especially critical in shaping new work habits required to achieve a safer workplace.

The serious-incident prevention process model utilizes measurement and feedback systems as the basis for identifying appropriate reinforcement milestones and to provide early warning of a need for corrective actions. Adding positive reinforcement actions into the workplace helps ensure that employees feel genuinely appreciated when performance meets or exceeds expectations—a simple but powerful concept. Monitoring upstream performance indicators provides a further excellent opportunity to initiate preventive actions before serious incidents occur, rather than after the fact, as is so often the case when the employees have no reliable indication of upstream performance indicators.

It is an old adage that "what gets measured, gets done." Perhaps a more accurate statement is that "what gets measured and reinforced, gets done." The reinforcement and feedback element of the serious-incident prevention process helps ensure that outstanding performance is recognized and early intervention is initiated when the need for adjustments to the prevention process is indicated.

18.8 ELEMENT 8: IMPROVE AND UPDATE THE PROCESS

Most workplaces are ever-changing. Changes in equipment, facilities, personnel, materials, and additional factors continually impact most organizations and the specific actions required to sustain safe operations.

The serious-incident prevention process model recognizes the importance of ongoing changes within organizations and the need for a systematic approach to ensure that actions required for incident-free operations remain effective. As illustrated by Figure 18.4, the eight elements of the process functioning together provide the framework for achieving and sustaining the workplace conditions necessary to operate free from the catastrophic consequences that often result from serious incidents.

18.9 IMPLEMENTATION OF THE SERIOUS-INCIDENT PREVENTION PROCESS: PIPELINE OPERATIONS CASE STUDY

History confirms that the potential frequency of incidents involving the transport of hazardous materials in pipelines is inherently low. However, personnel involved in pipeline operations fully understand that, in the event of a pipeline incident, the potential for catastrophic consequences clearly exists. Thus, a challenge for personnel operating pipelines is to maintain diligence and constancy of purpose even when operations seem almost always to go smoothly.

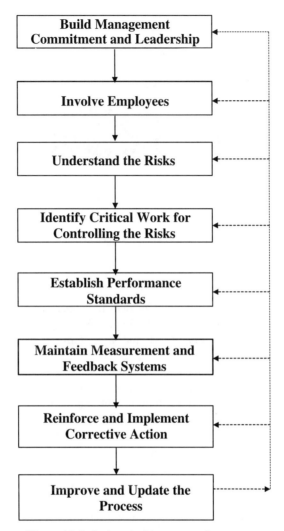

Figure 18.4. Serious-incident prevention process.

One company's pipeline team, comprised of a first-level supervisor and other personnel responsible for operating and maintaining the pipeline system, established an objective of developing a more effective approach to serious-incident prevention based on the eight-element process model. The team's assessment of risks included evaluating causes of past U.S. pipeline accidents. Department of Transportation (DOT) data (Fig. 18.5) identify third-party damage as the most frequent cause of pipeline incidents, followed by defective equipment or repair, external corrosion, internal corrosion, and operator error. The team's thorough evaluation of potential risks has helped ensure that the appropriate tasks and safe

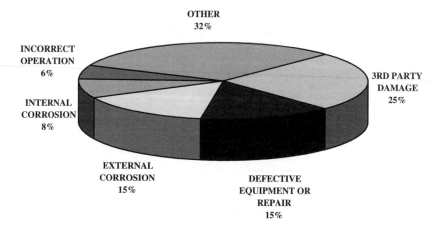

Figure 18.5. Causes of U.S. hazardous liquid pipeline accidents (based on year 2000 reports to DOT).

practices are identified for sustaining incident-free operations (Table 18.1). The pipeline team has established appropriate frequencies for performing tasks critical to sustaining incident-free operations and a system to document scheduled completion dates for the critical tasks.

The team monitors the percentage of critical work completed and charts the results on a monthly basis (Fig. 18.6). The chart includes each month's performance score, a 12-month moving average helpful in highlighting trends, and an indication of the team's improvement goal. Feedback, reinforcement, and corrective actions are often facilitated through the addition of handwritten notes on the charts that are posted in the work area to reinforce excellent performance and to pinpoint improvement needs.

The company's pipeline operations are subject to an annual regulatory agency audit to determine compliance with DOT requirements. Before implementing the improved management process, the team routinely received notices of violations following these audits. Since then, diligent execution of safety tasks identified as critical by the pipeline team has eliminated violation notices after the regulatory audits. Specifically, after implementing the improved process, the team completed five consecutive years with *zero* regulatory violations despite the continuation of thorough annual regulatory agency audits. The team's breakthrough improvement is reflected by the team's measurement system for monitoring the number of violations (Fig. 18.7).

The status of each performance measurement is reviewed by the pipeline team each month. Positive reinforcement is triggered when the measurements demonstrate that performance milestones have been achieved. When measurements indicate improvement opportunities, root causes are identified and proactive corrective actions are initiated. The team drives continual improvement by focusing on improvement opportunities controllable by the team.

TABLE 18.1. Serious-Incident Prevention Critical Work

Critical Work	Frequency
Third-Party Damage Prevention	
(a) Investigate each planned excavation near right-of-way	Prior to excavation
(b) On-site monitoring of all excavation work on right-of-way	Each excavation
(c) Fixed-wing aircraft aerial right-of-way patrol	Weekly
(d) Inspection to ensure right-of-way marker signs in place	Monthly
(e) Mowing of right-of-way	Each June and August
(f) Ground-level inspection of entire right-of-way	Annual
(g) Navigable waterway inspections by diver	Every 5 years
Defective Equipment and Repair Prevention	
Audit: <ul><li>Management-of-change process</li><li>Welding certifications for authorized repair personnel</li><li>Contractor safety and training programs</li></ul>	Semiannual audit
External Corrosion Prevention	
(a) Cathodic protection rectifier inspections	Monthly
(b) Interference bond inspections	Monthly
(c) Corrosion grid inspection	Semiannual
(d) Cathodic protection test station survey	Annual
(e) Close interval cathodic protection survey	20% of pipeline each year
Internal Corrosion Prevention	
(a) Monitor rate of corrosion inhibitor injection	Daily
(b) Corrosion coupon inspections	Quarterly
(c) Piping grid inspection	Semiannual
Prevention of Operator Errors	
(a) Review and update operating manual	Annual
(b) Formal pipeline operator training	Initial training within 1 month of employment; refresher every 3 years and prior to significant changes
(c) Procedure updates for process changes	Prior to implementing each change
(d) Review and update training requirements	Annual
(e) Ergonomic/human factors review of control systems	Annual
(f) High-pressure shutdown checks and instrument calibrations	Semiannual
(g) Audit of antidrug and alcohol misuse program	Annual

TABLE 18.1 (*Continued*)

Critical Work	Frequency
Emergency Preparedness	
(a) Test remote valve operation	Monthly
(b) Test low-pressure valve shutdowns	Monthly
(c) Exercise and inspect manual valves	Semiannual
(d) Visit public emergency response agencies	Annual
(e) Conduct emergency drill	Annual
(f) Test combustible gas analyzers at pump stations	Quarterly
(g) Inspect fire extinguishers	Quarterly
(h) Inspect and test uninterruptible power supply units	Quarterly
(i) Inspect and test relief valves	Per documented schedule
Other	
(a) Investigate all near misses	Each near-miss incident
(b) Audit safety permit system compliance	Semiannual
(c) Conduct safety review meeting with contractors	Semiannual
(d) Process hazards analysis	Every 3 years
(e) Confirm pipeline integrity with pressure test or smart pig	Every 10 years
(f) Review and update risk assessment process	Annual

The pipeline team's reinforcement plan (Table 18.2) includes primarily intangible reinforcement such as notes of appreciation and verbal recognition. Reinforcement milestones have been established both for achieving results and for improvement actions. Celebrations are also included in the reinforcement plan and have proven to be a further effective form of reinforcement for recognizing the achievement of major milestones.

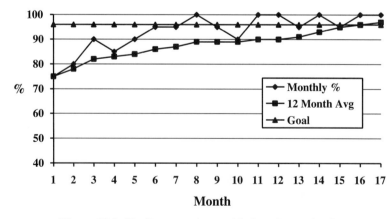

Figure 18.6. Pipeline operations: critical work completed.

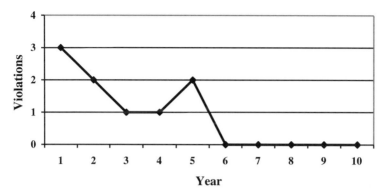

Figure 18.7. Pipeline regulatory agency violations.

The pipeline team also promotes the timely reporting of near misses and the investigation of other possible early warning signals of potential pipeline problems. Red-flag conditions, such as the prospect of future construction near a pipeline right-of-way or instrumentation not providing accurate readings, are proactively identified and investigated. Emphasis is placed on taking actions in the early stages to prevent the progressive development of more significant problems.

The team understands the importance of continually improving the process and is active in gathering and analyzing new information that has potential impact on

TABLE 18.2. Reinforcement Plan Pipeline Operations Serious-Incident Prevention

Result or Action to Reinforce	Who Receives Reinforcement?	Who Delivers?
Serious-incident prevention process fully developed	Pipeline team	First-level supervisor and department head
Performance measure fully implemented for 1 month	Process steward and other team members	First-level supervisor
100% completion of critical work achieved for month	Pipeline team	First-level supervisor
Completion of 12 months without a regulatory agency violation	Pipeline team	Chemical-handling department team
Action taken to identify and arrange for correction of red-flag condition	Individual or team taking the action	Chemical-handling department team (in form of red-flag note of appreciation)
Moving average above goal for 3 consecutive months and new goal established	Pipeline team	Chemical-handling department team

pipeline serious-incident prevention. The process is periodically reviewed and updated. The team is proud of its achievements and remains fully committed to its critical objective of maintaining incident-free operations. The serious-incident prevention process has provided major benefits for all stakeholders—employees and their families, managers, shareholders, customers, suppliers, regulatory agencies, and members of the public who live and work along the pipeline rights-of-way.

19 Special Topics: Self-Observation Process

A behavioral safety process based on self-observations may be more appropriate for some companies. A self-observation process is particularly appropriate when (1) their employees work in isolation or on small crews of two or three (such as loggers, electrical linemen, or truck drivers) or (2) their employees are extremely resistant to the idea of peer observations. In these situations, a self-observation process has been shown to improve safety practices and reduce incidents [see Olson and Austin (2002) and the case studies in Chapter 25].

Implementing a self-observation process is generally easier than implementing a peer observation process. With self-observations, observers do not have to be trained to provide feedback to their co-workers and do not need to develop skills in handling resistance to the observation process. In addition, the self-observation forms typically do not include comments related to conditions, and the steering committee usually does not have to do the same level of detailed analysis and problem solving that are required in a peer observation process.

A self-observation process addresses three primary questions that directly influence safe performance:

1. What's my job?
2. How am I doing?
3. What's in it for me?

A sports analogy illustrates how these three questions apply. Imagine for a moment you are a first baseman in the game of baseball. Is it clear what you need to do to do your job successfully? Yes, it is. The best behaviors and practices you would engage in are straightforward regarding fielding a specific area, holding runners on base, handling bunts and relays, and so on, in addition to hitting the ball when at bat.

A good deal of data-based feedback also exists for answering the second question both for the team and for individual players, not only in the game score and the team's won–lost record but in statistics regarding the team's and each player's performance.

Note: By Ann Pinney of Quality Safety Edge.

The answer to the third question is also clear in this example. For instance, when a first baseman executes a double play by catching a line drive and then tagging out a runner who has left first base, he receives accolades from his team and the fans, not to mention considerable self-satisfaction. Furthermore, in professional baseball, the player's financial success is closely tied to his performance.

The same three questions are addressed in the self-observation process and correspond to the stages of implementation.

19.1 WHAT'S MY JOB?

The individual worker or the crew members list the actions that must be engaged in to do the job safely. This list of actions or behaviors becomes the foundation for a detailed personal index of safe behaviors.

19.2 HOW AM I DOING?

After the index has been constructed, individuals or crew members identify a sampling method of self-observations that will enable them to measure their performance against the index at a specific time. The samples are triggered at random times at which the workers measure themselves with regard to the safety of the actions in which they are engaged at that moment. For example, the index in Figure 19.1 was constructed by a crew of loggers for the specific activity of cutting down a tree at a logging site. A faller typically cuts down trees with a falling partner. Although they are working in the same work area, they must be at least two tree lengths apart. They are often able to hear each other's saws, but they do not have visual contact. The two workers typically contact each other every hour or so via mobile radios that each carries. About three times a day, a faller will contact his falling partner at a random time and ask him to conduct a self-observation using the index above. At the end of the day all sampling observations are collected without names attached and calculated to create a team score.

Conditions/Behaviors	Yes	No	N/A
1. Boots are in good condition with sharp caulks			
2. Well-marked trail to work site			
3. Saw or axe ready at hand at the tree			
4. Overhead hazards checked			
5. Ground clear of possible hazards or chain reactions			
6. Getaway trail established			
7. Partner is at least two tree lengths away			
8. Clean undercut pertains			
9. Backcut is 2 inches above undercut			
10. Wedge is in backcut			
11. Body position is clear of hazards while bucking			
TOTALS: # Yes ÷ Total = % Safe Behaviors			
	Yes+	No	=Total

Figure 19.1. Example of a self-observation checklist (safety index) for loggers.

When the concept of such self-scoring is brought up, a common concern is that some employees may cheat on their score. Employees might inflate their scores initially, but they soon begin to see that the data will not be used against them, and they begin to trust the process. They realize that whether they check yes or no regarding performing safe behaviors, they have to think about what they are doing. They have had to lift themselves out of their more or less automatic work behavior long enough to assess their performance. It is through this assessment that they become more conscious of their safety practices and, as a result, can begin to work on improving their safe work habits.

When introducing this index and the self-monitoring process at one of the logging sites, a veteran faller commented, "I've been falling trees for thirty-five years, and since I still have all my fingers and toes, I don't think this process will help me." After trying it for two weeks, he reported with some surprise that at one point he would have had to put himself in danger to get to his axe. He added he never would have believed it had he not had to measure himself. That is typical for most of us; we are unconscious of our habits. When we are prompted by some external event, such as a safety observation, we realize what we are actually doing.

The main obstacle in the sampling process is not the time it takes to do a self-observation (most take less than 1 minute) but determining how to trigger samplings at remote work sites for isolated workers. Some methods for triggering them are discussed in the sections on index development and sampling later in this chapter.

19.3 WHAT'S IN IT FOR ME?

Individual crew members need to experience satisfaction and self-mastery as they improve in safety performance. This is actually a natural by-product when workers begin to conduct self-observations in earnest. Immediate self-correction starts to take place with every sampling. The challenges in using the sampling process are getting employees to try the process in the first place and next in keeping the process simple, meaningful, and fun to perform. Focusing on using external forms of acknowledgment along with celebrations for improvement and participation become critical in starting and maintaining the self-observation process. If it is a work situation in which employees rarely see one another, additional creative approaches must also be taken. These are addressed in the section on recognition later in this chapter.

19.4 HOW TO IMPLEMENT A SELF-OBSERVATION PROCESS?

Select a Safety Representative from Each Work Group

For both groups (which may include isolated solitary workers) and crews that work independently, the first step in installing the self-observation process is to have each group or crew select from their ranks a representative to learn the process in detail. The safety representative's role is to trigger the samplings of self-observations, calculate and post the daily scores, and deliver individual and team reinforcement for participation and improvements in performance. Following is a list of specific safety representative duties:

1. Calculate the scores as soon as possible after a self-observation sampling has been called for and post the data on a graph.
2. Look for any improvements in participation in the self-observation process as well as for improvements in safe behavior and provide recognition.
3. Manage the recognition budget.
4. Talk informally with workers about how the index process is working for them and note when the index seems to be getting stale and should be changed.
5. Once a week, show the graph and briefly discusses the most recent scores with the group or crew members in a safety meeting. (This interchange may take place with individual solitary workers by radio or telephone.)
6. Provide recognition and conduct celebrations when significant improvements in participation and/or safety performance have taken place.
7. Choose an alternate to maintain the process for periods when he or she is not going to be available.

Create an Index

Each individual worker or crew member creates a list of desired safety actions he or she needs to engage in to perform the job 100 percent safely. As in developing peer observation systems, these workers should analyze past safety data on injuries, accidents, and near misses to determine the behavioral content of the initial index. A prototype index can typically be created by a group of performers such as tree fallers or truck drivers who do the same job but perform it alone. Once the prototype index has been established for a particular job, the index can then be customized to incorporate special safety behaviors that may pertain for the tasks of individual workers. Each individual or group checks to ensure that the listed safe behaviors meet the pinpointing criteria (see the pinpointing guidelines in Chapter 21). An element of pinpointing that can be modified for the self-observation process pertains to listed items not normally observable by another person but that the individual can measure during self-observation. For example, in some operations employees have elected to include items such as "mind on task" on their index. Difficult for anyone else to assess, this item definitely helps address a worker's safety concern of having his or her mind wander while operating a dangerous piece of equipment.

After all workers have listed the safety behaviors they believe would be most valuable to improving their safety, the behaviors are entered onto an index of items each followed by columns for checking yes, no, or not applicable (N/A). When a

Behaviors	Yes	No	N/A
Loading:			
1. Wait for signal from loader, back in when safe to do so			
2. Drive slowly when backing and pulling out and hooking up reach			
3. Provide feedback to loader on stability of load			
4. Voice signal to second loader during binding up			
5. Check for overheight and overweight			
6. Give horn warning and check for second loader when leaving area			
7. Check load for stability			
In Transit:			
8. Perform walk around inspection (pretrip, load, mechanical, etc.)			
9. Drive to speed limits and road conditions			
10. Be prepared for the unexpected			
11. Frequently check mirrors			
12. Follow correctly (use three-second rule)			
Off-Loading:			
13. Check that PPE is in good condition			
14. Maintain three-point contact when ascending or descending truck			
15. Ensure no overhead hazards when loading			
Trailer:			
16. Walk around reload deck to ensure trailer loaded securely			

$$\frac{\text{\# Yes}}{\text{TOTAL}} = \% \text{ Safe Behaviors} \qquad \text{Yes} + \text{No} = \text{Total}$$

Figure 19.2. Example of a self-observation checklist for logging truck drivers.

sampling self-observation is triggered, the individual checks yes if engaged in the listed behavior, no if not engaged in the behavior but should have been, and N/A if the item is not necessary or pertinent at the moment the self-observation was triggered.

Figure 19.2 provides an example of an index for truckers loading and hauling cut trees from a logging site to an area for unloading and sorting. As you will note, this index includes a variety of safety behaviors for each activity involved in the trucker's daily work duties. When a trucker receives a radio call from the dispatcher to conduct an observation, he or she may be involved in loading, transit, or off-loading. The trucker simply checks off the behaviors listed in the activity in which he or she is engaged at that time.

Develop a Sampling Process

Once each worker has developed an index, it is time to begin collecting a baseline. The first data item on a specific self-observation is actually the true baseline when feedback intervention has begun. However, most groups try to collect five or six observations before posting a group baseline. They must first establish some prompt that will trigger the random observation samplings. Work situations that involve mobile phones, beepers, radios, even plant horns can be used to signal that it is time to make self-observations. Some work situations demand more creativity in prompting the process. In a nuclear plant in which self-observations were taken, a small, inexpensive timer triggered sampling with an alarm inserted in each of the work packets prior to each shift. Further, the timers were in Braille for blind persons; thus, it was difficult for the workers to know when the alarm would go off. An engineering group for a forestry department used the advent of a plane flying overhead to trigger filling out the safety indexes.

The key element in the process is triggering the observation at random moments to get the most accurate picture of a worker's safety habits. The more samples taken, the more accurate and complete the picture. The frequency of sampling must be balanced with the flow of work and workers' normal attention spans. The general guidelines offered by most work groups conducting self-observations is three to four samplings a week for the first two months of implementation and an average of two samplings a week after the baseline has been posted.

Post the Self-Observation Data

Once the samplings have been triggered and completed, the workers either drop off their indexes at a collection box or hand it into their safety representative for calculating the group or crew score. Again, the index does not have a name on it, only the identifying group (for individual workers) or crew name (for crew members) and the date. The safety representative adds all the yes's and no's from each of the indexes handed in and divides the number of yes's by the total yes's + no's to obtain the percentage of safe performance for that day.

What happens in the case of solitary workers who are not members of a group or crew? That occasionally is the case, yet it is usually possible to ascertain who each worker usually works with during the week or with which group he or she mostly identifies. When celebrating improvements in participation in the safety process or safety improvements, the group with which such workers have the most affinity to should be the one with which their individual data need to be combined. The other workers in this group are most likely to have the most influence on their daily work habits and would likewise be most likely to compliment their individual improvements. A sawmill using the self-observation process split its employees into crews that did not perform the same work but worked together at the same time and location. Each crew consisted of several production workers and an individual on-site maintenance employee. Although the safety behaviors on the separate indexes were different, the influence they had on each other remained high, especially in the area of reinforcement.

As to posting the data, many of a company's considerations with peer observation data pertain also to the self-observation process. The data should be posted in the respective group's or crew's work area for review only and updated weekly. The applicable safety representative incorporates discussion of trends in the current data, analyzes specific behaviors that are getting many no's, and helps determine when changes or new behavior items should be incorporated on the index. If the scores begin to achieve 90 to 100 percent safe for five or six samplings, this usually indicates that the behaviors listed have started to get to a high and steady rate. It would make sense at this point to develop a new index with safety behaviors that help address new concerns that arise in the increased awareness stimulated by the self-observations. In these instances, the safety representative might want to trigger self-observation sampling using the previous index on occasion, just to make certain that the safety practices established in the early sampling are still strong.

Provide Group and Individual Recognition

Providing positive reinforcement to individual workers for engaging in self-observations as well as for contributing to group improvement in safety scores is delivered primarily by the group or crew safety representative. Because conducting self-observations is entirely voluntary, acknowledgment and encouragement for increased levels of sampling participation are critical. In fact, when the safety representative posts the day's score, also displayed is the percentage of participation for the sampling for that score.

Figure 19.3 is an example of a graph of safety index scores from a sorting area within a timber operation. This graph illustrates the percentage of safe performance from 25 workers with entirely different indexes who have sampled themselves. The sorting area is where fallen trees are hauled, cut, graded, and bundled into what are called booms to be transported to market via waterways. It is typically a tightly spaced location in which large equipment must be maneuvered carefully around those who are working with the logs on the ground; thus it is an area fraught with danger that produces either a very closely knit working group or more commonly a somewhat fractious group of workers who do not always get along. Getting workers to adopt a process that calls for workers to admit to themselves, as well as record, when they perform unsafe actions is a major milestone.

As demonstrated by the baseline data, the scores are somewhat inflated. Before posting the baseline data, the workers had expressed a great deal of fear that the scores would be used against them. When the safety representative posted the baseline, he explained the scores and thanked the workers for participating. To acknowledge their participation, he went to a dock away from the work area, caught two large fish that he traded in a nearby town for several homemade pies, and brought them back to the work area to celebrate the workers' participation during a break. Interestingly enough, the next sampling reflected a lower safety score but a higher level of participation. The representative interpreted the changes as workers realizing the scores would not be used against them, so they became more honest. This has been a fairly common occurrence in various newly adopted self-observation systems. That is also why new safety representatives are taught to avoid any emphasis on collective graph scores for the first two months and instead to put their reinforcement efforts toward increasing participation.

The success of the safety representative's reinforcement attempts largely depends on his or her ability to customize social and tangible reinforcement to the workers in the group or crew. Because the safety reinforcement budget is usually small, representatives must be judicious about what forms of celebrations or small tangible reinforcers they choose. The perceived value of the acknowledgment is also heavily influenced by the level of informal leadership and respect the safety representative commands within the group or crew. Noting small improvements

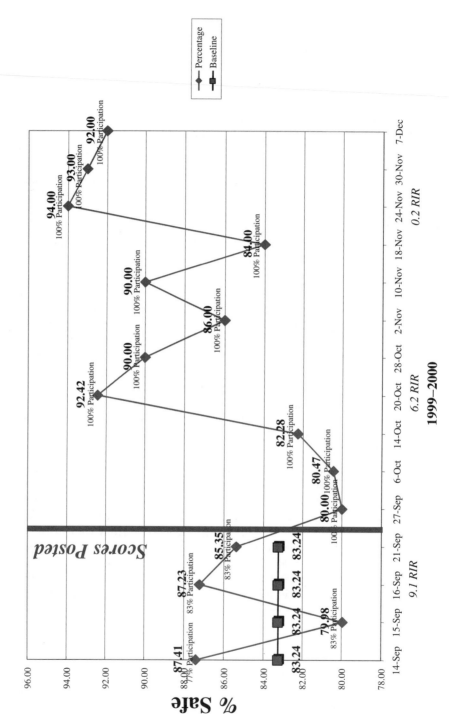

Figure 19.3. Data from a self-observation safety improvement process.

with positive comments and using innovative and fun ways to celebrate group improvement—as in the fish–pie instance—can make or break the early adoption of the self-observation process in a work site that is experiencing issues of low morale. We would also note that the group or crew members themselves select safety representatives, and once they begin to appreciate the power and value of the self-observation process, they sometimes replace an initial representative who was selected more or less casually or politically with an informal leader of the group or crew.

19.5 FINAL SUGGESTIONS ON SELF-OBSERVATIONS

Implementing a self-observation process is similar in many respects to a typical peer observation behavioral safety process but is much easier in other respects. The role of leadership is equally important in both approaches. While a self-observation approach often does not require the safety representatives or steering committee to do the same level of problem solving with the observation data, their use of observation data as the basis for reinforcement is even more critical in a self-observation process. If your organization opts for a self-observation process, these two elements warrant special study and attention.

20 Special Topics: The Steering Committee

This chapter provides additional details on the creation, training, and specific roles of the steering committee. After discussing formation and training, the remainder of the chapter describes how the steering committee should manage the process to maintain effective observations. Beyond the observations, the long-term success of a behavioral safety process requires that the steering committee analyze data and plan projects for improving safe work practices, which is the focus of the latter portion of this chapter and Chapter 22.

20.1 CREATING THE STEERING COMMITTEE

In the behavioral safety process, the initial employee-led design team usually becomes the steering committee after all employees have been trained as observers and the process has been kicked off. Steering committees are most effective with 5 to 8 members, although they may be larger to provide adequate representation. Design teams may be effective with 8 to 12; therefore, some method of decreasing the initial number of team members will be necessary. The effectiveness of these teams begins to decrease significantly as the number of members exceeds 12, and this number should not exceed 15. Actually, not all design team members may want to continue as members of the steering committee, whose term of office is usually one or two years, and so some members may volunteer to leave. If none do, you might ask some members of the design team to volunteer to serve on the steering committee at a later time.

In a large, task-diverse, or geographically dispersed organization, one central design team may be formed comprised of representatives of each work area that creates a template behavioral safety process for the organization. Following kickoff, each area or group of areas forms its own steering committee. In such cases these steering committees will need to recruit several additional employees to join the one or two representatives from their area who were on the design team. These steering committees next customize the central template for their specific area. They then manage their behavioral safety process as a separate process. Also, you should ensure that the team members' terms do not end at the same time. Once your

Note: This chapter was written by Grainne Matthews and Terry McSween, of Quality Safety Edge.

process is established, you should set up the schedule so that no more than two new design members start at the same time.

New steering committee members share the characteristics of the original design team members: nominated for serving on the committee by their peers because they are respected, natural leaders, interested in safety issues, and willing and available to devote several hours per month (often on their own time) to work on the process.

20.2 TRAINING THE STEERING COMMITTEE

In addition to learning how to manage the integrity, outcome, and results of the behavioral safety process, the steering committee members undertake training to learn how to

1. Identify, encourage, and reinforce safe work behaviors and practices by all employees and management that support the process;
2. Present safety data to other employees in meetings as feedback and lead employees in setting goals and choosing celebrations;
3. Assess whether the process is remaining true to the values identified by the design team and to the behavioral principles on which its success depends and, if it is not, raise their concerns to other committee members; and
4. Make effective use of time and resources in committee meetings.

These skills may be taught in an initial two-day workshop followed by review and feedback during subsequent steering committee meetings. In the case of organizations that have more than one steering committee, meeting together regularly to share best practices and problem-solving techniques can be valuable. Many steering committees also receive continuing education (and recognition) by attending safety conferences and seminars.

20.3 STEERING COMMITTEE RESPONSIBILITIES

Steering committees play a vital role in the success of the behavioral safety process. They are responsible for managing and improving three aspects of the process:

A. Process measures
B. Behavioral safety measures
C. Safety results or outcome measures

A. Managing Process Measures

The behavioral safety process consists of two major components based on five core principles. Successful behavioral safety processes are true to these principles. The steering committee protects the integrity of these principles.

1. **Component 1: Employee Involvement**

    (a) *Principle 1*: Employees who participate as observers improve their safety behaviors as much as or more than those they observe (see Chapter 27). Therefore, the more workers who conduct observations, the better.

    (b) *Principle 2*: The more often employees are observed and receive feedback, the more likely they are to improve their safety behavior. Therefore, the more often observations are made, the better.

    (c) *Principle 3*: Much of the value of the behavioral safety process is in exchanges between observers and those who are observed. Therefore, high-quality observation and feedback sessions are critical.

2. **Component 2: Data Analysis and Improvement Projects**

    Most of the remaining benefit of the behavioral safety process comes from improvements recommended or implemented by the steering committee based on its analyses of observation data and related information.

    (a) *Principle 4*: Therefore, the committee must conduct high-quality, timely analyses of the data and target behaviors.

    (b) *Principle 5*: Based on this analysis, the committee must also develop and implement action plans targeting improvement.

Steering Committee Functions. The following tasks are necessary for the committee to ensure the integrity of the behavioral safety process. These tasks can be grouped into specific roles or can be divided among the committee members. Their completion needs to be monitored to ensure that the steering committee remains on track:

- **Chair**

    Sets the agenda, chairs meetings, tracks attendance, and organizes logistics (meeting location, materials, refreshments).

- **Data Coordinator**

    Collects and provides observation checklists, data tables, and graphs; also arranges for additional information if necessary for analysis. For example, this individual may interview observers to clarify comments on their observation forms or may visit work locations to observe tasks discussed on observation forms.

- **Incident[1] Coordinator**

    Collects incident documents and arranges for additional information if necessary for analysis. For example, this person may interview people

---

[1] *Incident* as used here refers to any incident that has the potential to cause injury or illness. Thus it includes actual lost-workday injuries and illnesses, non-lost-workday injuries and illnesses, first-aid cases, near misses, vehicle incidents, and equipment damage incidents (such as forklift collisions with stationary objects, pump failures, spills, etc.). Some of these incidents may not be routinely documented in many organizations.

involved in an incident to clarify information in the documentation or may invite someone involved in the analysis to the steering committee meeting to elaborate.

- **Improvement Project Coordinator**

  Tracks action items on improvement projects. The coordinator may assign these items to steering committee members, managers and supervisors, or other employees.

- **Recognition and Celebration Coordinator**

  Plans celebrations with management when goals are met and schedules individual employee recognition by steering committee members, supervisors, co-workers, or others as designated (see the recognition and celebration guidelines in Chapter 22).

- **Scribe**

  Records decisions and publishes the minutes.

Each member can also function as an area representative for his or her department or team. Area representatives may do the following:

- Post area feedback graphs, tables, and lists on bulletin boards in their area.
- Personally encourage participation by their fellow employees.
- Deliver or arrange recognition as planned by the recognition and celebration coordinator.
- Report on the behavioral safety process to area supervisors and managers.
- Discuss the process, provide area feedback, and lead fellow employees in setting new goals at area meetings (see the section on setting goals and providing group feedback in Chapter 22).

Should the steering committee decide to create specific roles for committee members, there are several ways to set them up. Some committees assign the roles for the members' entire term of office, which is the simplest method. Other committees rotate the roles, which enables each member to learn about several aspects of the behavioral safety process. In the latter case, longer rotations of perhaps six months are generally best because they provide members with enough practice to become proficient at each of the various committee tasks.

Addressing Process Integrity. During its monthly meetings, the steering committee addresses the integrity of the behavioral safety process. It must monitor various measures and trends in each of the basic components. The steering committee should routinely ask the following questions to assess how well they are doing and help identify areas that may need attention:

1. **Component 1: Employee Involvement**
   (a) Have employees set their own reasonable improvement goals for each involvement measure?
   (b) Are employees receiving accurate, timely, and easy-to-understand feedback about progress toward their goals via publicly posted graphs and announcements in meetings?
   (c) Are celebrations happening as soon as possible after goals have been met, ideally within the week but no later than two weeks?
   (d) Are individuals receiving recognition as soon as possible after meeting recognition criteria, ideally within 24 hours but no later than three days?
   (e) If progress is not being made toward involvement goals, the steering committee conducts a behavioral analysis to determine what antecedent(s) or consequence(s) may need to be strengthened or added:
   - Percentage of participation: percentage of all trained observers who are conducting scheduled observations
   - Number of observations: total observations
   - Percentage of quality observations: percentage of observations containing pinpointed comments for the noted safety concerns and indicating which items and concerns were discussed

2. **Component 2: Data Analysis and Improvement Projects**
   General:
   (a) Is the committee meeting at least monthly?
   (b) Does it have at least 90 percent attendance at meetings?
   (c) Are its members completing agreed-upon functions between and during meetings (as described above)?
   (d) Do its members represent all departments and shifts?
   (e) Is it soliciting input from employees in each department and on each shift?
   (f) Is it analyzing data monthly to address two to three target practices?
   (g) Is it using information from observations, incident reports, experience, and interviews to conduct behavioral analyses of antecedents and consequences for target practices?
   (h) Are its improvement projects based on missing or weak antecedents or consequences?
   (i) Is it using information from incidents reports to do the following?
   - Improve the checklists—for example, improve definitions, add examples to definitions, or reorder checklist items
   - Guide the process—for example, encourage observations of tasks at times when injuries have occurred but few or no observations have been made

Improvement projects on targeted practices:

(a) Are employees setting reasonable new improvement goals when previous goals have been met?
(b) Are they receiving accurate, timely, and easy-to-understand feedback about progress toward goals via publicly posted graphs and announcements in meetings?
(c) Are celebrations happening as soon as possible after goals have been met, ideally within the week but no later than two weeks?
(d) Are individuals receiving recognition as soon as possible after pertinent behavior, ideally within 24 hours but no later than three days?
(e) Are observers focusing on targeted practices with observation and feedback?
(f) Are responsible employees implementing agreed-upon improvement plans?
(g) Are they receiving recognition when they do so?

The steering committee asks these questions at every meeting and makes plans to correct deficits where identified. A behavioral safety process will not achieve its full potential in terms of an increase in people working safely and a decrease in incidents if these components are not implemented as described.

B. Managing Behavioral Safety Measures

If the process is functioning the way it was designed, employees should be working more safely. The percentage of safe behaviors (i.e., the percentage of times an observed practice is marked as "safe" on the observation checklist by observers) should increase for all practices as a result of increases in observations and feedback. The percent safe on practices targeted by the steering committee in improvement projects should increase even more. Finally, work practices that have reached 95 to 100 percent safe need to remain at that level for an extended period (generally a year) to ensure that the improvement is not the result of natural variations over time.

Accordingly, the steering committee reviews observation data to ensure that the process achieves the desired outcomes:

1. What is the percent safe on each checklist practice?
2. Is the percent safe on practices targeted by improvement projects increasing?
3. Have some of the safe practices remained near 100 percent for a year or more?

C. Managing Safety Results or Outcome Measures

If people are working more safely, the organization should experience fewer injuries. The steering committee monitors the behavioral safety process data results

TABLE 20.1. Steering Committee Responsibilities

Instructions: For each measure, start at column 1 and work across to the last column. The responses to each question will indicate what steering committee actions are necessary. Shaded areas indicate that the action is not appropriate for that measure. *Note*: Not all measures are included.

Steering Committee Responsibility	Data/Measure	1. Is Graphic Feedback Posted And Current?	2. Has Goal Been Set (What) Or Met (Date)?	3. If Goal Has Been Met, Has Employee-Chosen Celebration Been Arranged?	4. If Anyone Has Earned Individual Recognition Based On Plan Criteria, Has It Been Arranged?	5. If Goal Has Not Been Met, Is Improvement Project Needed?	6. Is Any Change In The Process Needed?
Manage integrity	PARTICIPATION						
	(a) % Participation						
	(b) No. observations						
	(c) % Quality observations						
	STEERING COMMITTEE						
	(a) % Scheduled meetings held						
	(b) % Attendance						
	(c) % Responsibilities completed						
	(d) % Departments/ shifts represented						

	IMPROVEMENT PROJECTS						
	(e) % Observations that include targets						
	(f) % Improvement actions completed						
Manage behaviors	1. % Safe on all checklist practices						
	2A. % Safe on target practice 1						
	2B. % Safe on target practice 2						
	2C. % Safe on target practice 3						
	3. % Met target practices still at 95–100%						
Manage results	No. of incidents related to practices on checklists						

to ensure that the process is achieving this ultimate goal. In addition, the steering committee uses this information to improve the process.

1. If an incident occurs, the steering committee pinpoints the safe behavior that could have prevented it or lessened the severity of the incident. (Pinpointing is discussed in more detail in Chapter 21.)
2. If the practice is on the checklist, the steering committee determines if the frequency and severity of incidents related to this practice have been decreasing over time. If they have been, does this incident provide any new information that would merit any changes in the process?
    (a) If incidents have not been decreasing, the steering committee determines whether the percent safe for this practice has been increasing over time. If it has, does this incident provide any new information that would merit any changes in the process?
    (b) If the percent safe has not been increasing, the steering committee asks:
        (i) Is the practice being observed often enough (number of observations that include this practice)? Are enough of those observations high quality?
        (ii) If the practice has entailed an active improvement project, is the project based on a thorough and accurate behavioral analysis? Does this incident add any new information to the analysis? Is the project occurring as planned?
3. If the practice that could have prevented or reduced the security of the incident is not on the checklist, does it occur frequently enough or is the risk high enough that it should be added?
4. Also, if the behavior is on the checklist, the steering committee should consider whether the task that was being performed when the incident occurred was one that is routinely observed. If not, then it needs to develop plans to increase the frequency of observations of this particular task.

20.4 RESPONSIBILITY SUMMARY

Table 20.1 summarizes how the steering committee manages the behavioral safety process. It can be used as a guide for conducting regular steering committee meetings. It can also be used to document the results of each meeting by filling in the answers to each question for each measure or by checking that that item was addressed. The steering committee meeting agenda in Figure 20.1 outlines the same activities in a different format.

Steering Committee Meeting Agenda

Date:_____ Chair:_____ Scribe:_____

Team Members Present	
Team Members Absent (and reason)	

Order of Business
1. **Review:**
   - Mission
   - Values
   - Meeting Ground Rules

2. **Process Measures**
   - Percent Participation
   - Number of Observations
   - Quality Observations
   - Steering Committee Functioning
   - Improvement Projects

3. **Behavior Measures**
   - Pareto* % Safe
   - Pareto Number of Concerns
   - Pareto % Concerns
   - Trend % Safe on Target Practices
   - Incident Documentation
   - ABC Analysis

4. **Results/Outcome Measures**
   - Incidents Related to Checklist and Targeted Practices
   - Severity of Injuries (lost workday, recordable, first aid)
   - Injuries *Not* Related to Practices on Safety Checklists

Assignments for Next Meeting

Who	Will Do What	By When

*Refers to bar charts on which the bars appear in descending (or varying) order of magnitude enabling quick identification of the most frequently occurring items.

Figure 20.1. Sample agenda form for the steering committee.

21 Advanced Topics: Why It Works and Behavioral Basics

We have talked about behavior throughout this book, but what are we talking about? What is behavior? Furthermore, why do people behave the way they do? These are two important questions, particularly for those interested in improving safety.

A basic understanding of the theory that underlies behavioral technology will help you make your efforts successful. Such an understanding will be useful as you (1) develop your implementation plans, (2) problem solve and refine your efforts, and (3) analyze incidents to plan preventive measures. The process is straightforward in theory but often is more difficult in practice. First you specify the desired behavior; then you analyze the situational and historical events that support a particular behavior.

We will start with the first question: What is behavior? Behavior is simply anything someone does or says. Behavior is any activity that a dead person cannot do, any muscular or glandular action or reaction (Malott et al., 2000).

Behavior is not personality, attitude, or intelligence. These are labels or abstractions that we often use to describe someone's behavior in some specific way. For purposes of this book, I have used the term behavior only for observable actions, because observable practices are what concern us at work. Accordingly, I have talked about specific observable acts.

While attitudes may be important, the behavioral approach addresses how people behave on the job. We can only know someone's attitude by our observations of how they behave and what they tell us. If we can change their safety habits, their attitudes about safety will follow, especially as their colleagues also adopt better safety habits. Once we have a group of people with similar habits and attitudes about safety, then we begin to talk about people having a common safety culture. Thus when we want to talk about changing the culture, we have to talk about changing people's behavior. When we get behavior change, changes in attitude and culture will follow.

21.1 PINPOINTING

Pinpointing is describing behavior in observable terms, without the use of labels or abstractions. It means simply being specific in describing exactly what someone does without interpreting or evaluating the observed behavior. Pinpointing means avoiding the use of labels and objectively describing the actual behavior that might lead someone to use a label in normal conversation.

The first step in designing a behavioral safety process is pinpointing a list of behaviors critical to safety without the use of labels or explanations of behavior. Complex safety practices must be broken into components and defined in precise detail.

Sometimes, you need to identify the results of behavior rather than the actual behavior. Doing so has an advantage in that the results remain when the behavior is no longer occurring. For example, you can observe spilled oil without having seen someone spill the oil (or the failure to clean it up). Figure 21.1 presents several examples of common labels, pinpointed behaviors, and results.

Pinpointing is often a two-step process:

Step Number	Activity
1	Pinpoint the problem, undesirable behavior (i.e., an unsafe act), or undesirable result of the behavior (an unsafe condition).
2	Pinpoint the desired behavior, what you want employees to do (i.e., a safe act), or the desirable result of the behavior (a safe condition).

The reason for identifying the desired behavior is that it allows you to focus on the positive. Once you identify the desired behavior, you can plan ways to encourage and support the desired behavior. Pinpointing the undesired behavior also enables you to plan ways of discouraging such behaviors, but this is the traditional and less effective approach to safety. Establishing a preventative safety initiative requires that you know what behaviors you want to encourage and that you design an environment that provides positive support for those behaviors.

Labels	Behaviors	Results
Aggressive	Filling out a checklist correctly	Area marked off with safety tape.
Angry		
Bad attitude	Proper lifting (legs bent, back straight)	Electrical wires laying across walkways
Brown-noser		
Dangerous	Putting lock-out tags on electrical switches	Lock-out tags in place
Good worker		Rate of OSHA-recordable incidents
Neat	Wearing safety glasses	
Unsafe	Wearing safety harness when working above six feet	

Figure 21.1. Examples of common labels, behaviors, and results of behavior.

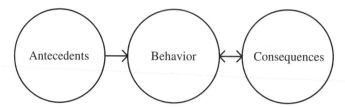

Figure 21.2. Relationship between antecedents, behavior, and consequences (preliminary diagram).

21.2 ABC ANALYSIS

Our basic premise is that behavior is a function of the immediate environment. Once we have pinpointed a specific behavior, we can then divide environmental events into two sets of categories: events that precede the behavior and events that follow the behavior. Behavioral psychologists use the terms antecedents for events that occur before the behavior and consequences for those that follow behavior. Figure 21.2 shows a diagram representing these events. Antecedents are events that precede behavior and prompt or cue the occurrence of that behavior. Consequences are events that follow behavior and that influence the likelihood that the behavior will occur again under those antecedent conditions in the future. Consequences either strengthen or weaken behavior.

Notice the two-way arrow between consequences and behavior in Figure 21.2. It suggests that consequences affect the likelihood that the behavior will occur again. They may either strengthen or weaken the behavior. Consequences cause the frequency of behavior to increase or decrease. In other words, they can increase the likelihood that the behavior will occur again under similar conditions or they can decrease the likelihood that the behavior will occur again under similar conditions.

The relationship between these behavioral events is a contingency relationship, that is, an if–then relationship. If the antecedent conditions are present, then the behavior will occur. If the behavior occurs, it will be followed by the consequence.

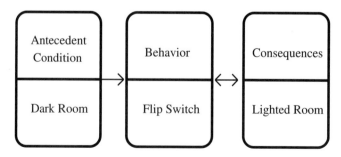

Figure 21.3. Contingency diagram of the environmental events associated with flipping a light switch.

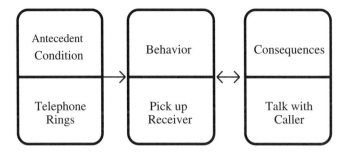

Figure 21.4. Contingency diagram of the environmental events associated with answering the telephone.

Let us now look at some examples. If you walk into a dark room, the dark room is likely to be an antecedent for what? It will usually prompt you to flip or look for a light switch. The consequence of flipping the light switch is what? When you flip a light switch, the light comes on and you can find your way around the room.

We might diagram this event as shown in Figure 21.3. This diagram is called a contingency diagram in showing the if–then relationship of the primary behavioral events (Malott, 1992a,b).

Figure 21.4 presents another example.

In each of these examples, what is controlling the behavior? The answer is that both the antecedents and the consequences are controlling the occurrence of the behavior. The antecedent prompts you to respond. In a sense, the antecedent condition determines when the response occurs. Still, you would not continue to flip the switch if the light never came on. Nor would you continue to answer the phone if no one responded. Consequences are important in maintaining the effect of the antecedent condition. In fact, the antecedents are effective in prompting behavior only because of the consequences. Since the antecedent would not be effective without consequences, the consequences are what really control behavior.

But why is the behavior occurring in the first place? The behavior occurs because of a history of experience with these antecedents and consequences. Sometime in the past, we learned about light switches and telephones. We will add "individual learning history" to our diagram after we further discuss consequences and antecedents.

21.3 CONSEQUENCES

As mentioned, consequences can either increase or decrease the behaviors they follow. The two types of consequences are generally familiar: Reinforcement increases behavior while punishment decreases behavior (see Fig. 21.5). In addition, behavior may be followed by a neutral event that neither strengthens nor weakens the behavior. As a rule, safety improvement efforts need to add consequences that support safe behavior on the job. The consequences occurring in the natural work environment simply do not maintain the levels of safety that we

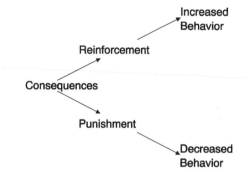

Figure 21.5. Primary types of consequences and their impact on behavior.

strive for in today's workplace. The key to improving safety performance is identifying and arranging consequences that support compliance with safety procedures.

Built-in Consequences. The probability of getting hurt because of not following a safety procedure is part of the consequences "built in" to the job itself. Many consequences involved in safety are natural consequences that occur simply as a result of engaging in the behavior. Discomfort associated with wearing protective equipment on a hot day is built in to the act of wearing the equipment and makes the use of that equipment less likely. Following safety procedures often takes more time than short cutting those procedures. Climbing a structure without a safety harness or making a quick weld without a hot-work permit may result in getting the job done more quickly. Conversely, if a tool is dropped from above, the head is protected by a hard hat. The protection from injury is reinforcement that is built in to the act of wearing the hard hat.

Unfortunately, most built-in consequences support unsafe acts rather than compliance with safety procedures. Too often, following safety procedures creates delay or discomfort or inconvenience that punishes compliance with established safety practices. However, following safety procedures reduces the risk of injury, which is also a built-in consequence. Unfortunately, the likelihood of actually avoiding injury by following the safety procedure is usually too improbable to provide reinforcement significant enough to offset the built-in punishing consequences.

Added Consequences. Because most of the built-in consequences do not support safe practices, we must arrange contrived consequences in the workplace. We use social consequences to provide such support. Social consequences require that another person be present and act immediately following someone's behavior.

In traditional safety management, the most common added consequence is the threat of punishment. It can take the form of corrective feedback, criticism, nagging, or disciplinary action. Employees follow the safety procedures to escape from or avoid punishment.

In contrast, the behavioral approach provides added positive consequences for safe behavior. Added reinforcing consequences include positive attention from management and from peers. Such attention may include simple personal praise as well as suggestions or offers of support. Added reinforcement may be provided in the form of publicized comments that employees are making progress toward a goal or have earned recognition or awards. When we create an observation process, the intent is to add such social consequences to support safe behavior on the job.

The behavioral safety process also provides additional corrective feedback from observers. But the observers are more likely to be peers and the corrective feedback does not have the same threat of performance evaluation included in the traditional approach to safety.

Delayed Outcomes. We often create safety award programs to encourage employees to behave safely. The awards usually provide additional positive outcomes for not getting hurt but are too delayed to be effective consequences. We discussed problems with such programs earlier, yet such programs can be an effective component of a safety process. Figure 21.6 displays a more complete diagram that includes both consequences and outcomes.

Self-Provided Consequences. The ideal safety program is one that maximizes self-control. The employees know the safety rules and feel good about behaving in ways consistent with those rules, often in spite of built-in consequences. They feel anxiety when they take shortcuts or fail to follow safety rules. People who consistently wear seat belts feel similar discomfort if they ride in an automobile without their seat belt fastened. Such "self-control" is the goal of added consequences and adding delayed outcomes such as safety awards. The "anxiety" results from the potential punishment for breaking safety rules and the potential loss of future safety awards, compensation, and advancement opportunities.

Added consequences and outcomes help create and support self-provided consequences. Seat belt laws provide added consequences that help increase the self-provided consequences that have increased seat belt use. A goal of the behavioral safety observation process is to add social consequences that support self-provided consequences. A complete behavioral process further includes

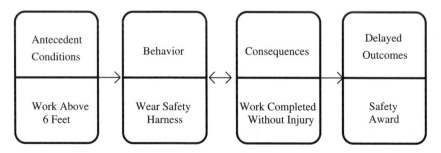

Figure 21.6. Contingency diagram adding delayed outcomes.

establishing team improvement goals, safety award programs, and other programs that support self-provided consequences for safe behavior.

Factors Influencing Effectiveness of Consequences. Two factors influence the effectiveness of consequences and delayed outcomes:

- Probability of the consequence or outcome
- Significance to the individual

Often, safety award programs do not affect behavior on the job because (1) the awards are not significant to the employees or (2) the employees are likely to get the award regardless of safe or unsafe practices on the job. Some safety programs continue to use awards and forms of recognition that are not meaningful to employees because they already have more coffee mugs, baseball caps, and other trinkets than they will ever need. One of the challenges in designing an effective safety award program is to be able to select awards that will be meaningful to particular groups or individuals.

These factors help explain why people do not comply with safety procedures simply to avoid getting hurt, even when they clearly know the correct procedures. The probability of getting hurt from failing to comply with a given safety procedure is generally too small to be effective in maintaining safe practices. Given the incidence rates for most industries, the probability of being hurt on a given job is fairly low, even if an employee fails to comply with safety procedures every time. If the incidence rate is 5 per 200,000 work-hours, for example, an employee will have an injury based on chance about once every 20 years. These probabilities are part of the problem.

If the risks were greater, employees would be more likely to comply with safety procedures. If employees had a 1 in 4 chance of a serious fall every time they worked above 6 feet, ensuring use of a safety harness would be no problem. If employees had a 1 in 3 chance of losing their eyesight any time they worked without safety glasses, ensuring the use of eye protection would be no problem. Fortunately or unfortunately, most organizations are safe enough that employees can be complacent. Our challenge is to provide added consequences that support safe work practices.

21.4 ANTECEDENTS

An antecedent must be present immediately before the behavior of interest. When identifying an antecedent, you should identify the stimulus conditions that prompted the behavior to occur. Safety antecedents include verbal instructions, signs, and the situation that prompted action.

Other events not immediately preceding the behavior of interest may include written procedures, safety rules, and safety improvement goals. Because these events do not immediately precede safe behavior, you should consider them part of the person's history rather than an antecedent. Often, an individual's unsafe response to a situation may suggest a problem with the training or instruction that the employee has received some time prior to the behavior of interest. These elements will be discussed in greater detail in the next section.

A written procedure may be an antecedent if the employee referred to the procedure immediately before starting the job. Otherwise, the procedure and training on that procedure are part of the employee's individual learning history.

Key point: Antecedents affect behavior because of the consequences. In the contingency diagrams regarding the light switch and the ringing telephone, the antecedents prompted the behavior to occur because of the consequences that usually follow in the presence of those antecedent conditions.

21.5 INDIVIDUAL LEARNING HISTORY

As mentioned, many behavioral events, such as training, role models, and reviewing procedures, occur too far in advance of the behavior to be considered antecedents yet are important to how a person responds in a given situation. These factors often establish employees' skills or knowledge of job procedures, both actual on-the-job practices and formal procedures. In formal training programs employees generally learn what they are supposed to do. Then, through on-the-job contact with more experienced employees and their supervisors, they learn how the jobs are actually done. Sometimes the procedures and on-the-job practices are the same. Sometimes there is a difference between the two. Figure 21.7 presents an expanded diagram that includes learning history.

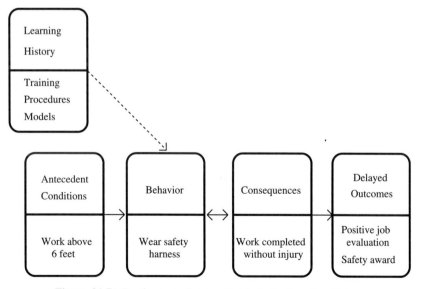

Figure 21.7. Contingency diagram that includes learning history.

This expanded picture of the behavioral environment clarifies all of the functions of the behavioral safety process. This safety process supports safe practices on the job by:

1. arranging immediate social consequences that support safe practices,
2. providing training (a learning history) that sharpens the observer's understanding of what is and what is not safe under different antecedent conditions, and
3. supplying delayed outcomes (team celebrations and individual recognition) that support both safe practices and the observation process.

The observation process provides social consequences for safe practices through the feedback that observers provide to those they observe. The positive feedback provided immediately following observations of safe practices should reinforce those practices, while corrective feedback should prompt safe practices and decrease the likelihood of future unsafe practices.

In addition to the consequences, identifying safe and unsafe practices during the observation process trains employees to identify risks in their daily work activities. In many ways, the observation process is analogous to arranging for employees to watch videos and identify safe and unsafe practices and conditions. Employees gain a clearer understanding of what they can do to create a safer work environment. The understanding and new skills help employees ensure a high level of safe practices all of the time, not only when they are conducting formal observations.

Finally, the behavioral safety process provides delayed outcomes that also support safety. The observation data can be the basis for both team celebrations and individual recognition. By participating in a formal observation process, managers and supervisors can more accurately evaluate employee safety performance when making decisions regarding compensation and promotions. Thus the process helps the normal organizational contingencies to support safety more effectively.

In addition, managers and supervisors may be evaluated on how well they implement and maintain the observation and feedback processes. The delayed outcomes (such as the manager's annual performance appraisal) then also support the observational feedback process that has a more direct impact on all employees' daily safety practices.

21.6 BEHAVIORAL ANALYSIS WORKSHEET

Analyzing behavioral incidents is a difficult task. Too many safety investigations stop short with a simplistic analysis, such as "failure to follow procedure" or "operator inattention," without doing a thorough analysis of the environmental factors that actually contributed to the behavior. The worksheet in Figure 21.8 provides an initial overview of factors that may have contributed to an unsafe act. These concepts will be developed more fully in the next chapter.

	Conceptual questions	Specific Questions
1. Behavior	What was the undesired behavior?	What was the unsafe act that resulted in an incident?
	What is the desired behavior?	What was the safe act that could have prevented the incident?
2. Consequences	What built-in consequences support the desired and undesired behavior? (Consider both reinforcement and punishment.)	Are the employees more comfortable or able to complete the job more quickly or easily by engaging in the unsafe act? Are employees uncomfortable or less productive if they do the job safely? Is it more work?
	What added social consequences support the desired and undesired behavior? (Consider both reinforcement and punishment.)	Do supervisors and others routinely provide corrective action for employees who engage in the unsafe act? Do supervisors and others routinely provide positive feedback for the safe act?
3. Delayed Outcomes	What delayed outcomes should support the desired behavior? (Consider both reinforcement and punishment.)	What safety awards should support the safe act? What effect does performing the unsafe act have on an employee's compensation or career?
	How significant are the outcomes?	Are the awards significant to employees?
	How probable are the outcomes?	Is it likely that an employee will lose a safety award for performing this job unsafely? Or earn a safety award for performing this job safely? Is it likely that performing this job safely or unsafely will affect an employee's compensation or career opportunities?
4. Antecedents	What antecedent condition prompted the unsafe act?	Was the employee encouraged to be in a hurry? Was the employee instructed to do the job incorrectly?
	What antecedents should have prompted the safe act?	Did the location of the incident have signs to prompt appropriate procedure? Were the correct tools and materials present to do the job correctly? Did the employee have a checklist at hand (or should one have been provided or required)?
5. Individual Learning History	What other events occurred in the individual's history that may have affected the likelihood of the safe or unsafe behavior?	Are the procedures correct? Did the employee know how to do the job correctly (and demonstrate mastery of the procedure during training)? Do other employees routinely model the correct way to do the job? Do employees know how engaging in the unsafe act affects their personal safety, their safety awards, and their careers?

Figure 21.8. Preliminary worksheet to assist in analyzing incidents and behaviors.

If:	Then you might:
The job procedure or equipment could be redesigned to eliminate the risk	Redesign the equipment or procedure
The safe practice occurs in specific work areas and signage is nonexistent or weak	Add or improve specificity of signs
The safe practice involves a skill or knowledge that is used regularly but the employees do not have	Arrange for job-specific training that provides the needed skills
The safe practice involves complex, infrequently used skills or knowledge that the employees do not have	Develop and require use of a checklist for tasks Train personnel to complete the tasks using the checklist
The built-in consequences support the desired behavior	Redesign the job so that doing the task safely is more comfortable and easier and takes less time
The safe practice is not already on the behavioral observation checklist	Add behavior to the observation checklist
This particular safety practice is not explicitly included in the definitions on the back of the observation sheet	Add a pinpointed example to the applicable definition
Employee had not been observed recently because the job function was not routinely included in observations	Revise procedure to ensure all employees are observed
Employee had not been observed recently because scheduled observations are not occurring consistently	Enhance procedure to improve completion of scheduled observations

Figure 21.9. Preliminary guide to assist in developing action after analyzing the target behavior.

21.7 DEVELOPING AN ACTION PLAN TO ADDRESS BEHAVIORAL CAUSES

Once you have analyzed the situation by considering the questions in Figure 21.8, use the guide in Figure 21.9 as a preliminary guide for developing action plans. This topic will also be developed in more detail in the next chapter.

22 Advanced Topics: Improvement Projects

This chapter outlines the data that the steering committee analyzes and the steps it takes to select practices for improvement projects. Steering committees are most effective when focused on no more than two or three projects at a time. A suggested strategy is to gather baseline data for one reporting period, typically one or two months, then focus on the one most important practice needing improvement. Once this first improvement project is underway, perhaps two months later, the second project is selected. The third improvement project might be planned two months after that. In this way, the steering committee is not overwhelmed, yet is seen to be taking action. Also, more data are made available for each subsequent analysis.

During each improvement project, the steering committee monitors the implementation of its recommendations and the percent safe on the target behavior. If the percent safe is not improving, the committee refines its recommendations by reviewing and repeating its analysis perhaps with new or better information. When the goal of 95 to 100 percent safe is met for several consecutive reporting periods (the exact number of periods depends on how frequently the practice targeted for improvement occurs), the committee decreases the amount of attention that it, the observers, and management pay to that practice and, using the same steps, selects another practice that is causing concern. The committee may decide to delete a targeted practice from the observation checklist when it has achieved 95 to 100 percent safe for a year or more or when revising the checklist.

22.1 PROBLEM-SOLVING STEPS

Step 1: Review the Pareto chart of percent safe on all checklist practices for the last reporting period. Which practices were less than 100 percent safe?

Step 2A: Review the Pareto chart of the number of concerns on practices that were less than 100 percent safe for the last reporting period. Which of these practices was observed to be unsafe most frequently?

Step 2B: Review the Pareto chart of the percentage of concerns on checklist practices that were less than 100 percent safe for the last reporting period. Which of these practices was observed to be unsafe a high percentage of the times it was observed?

Note: This chapter was written by Grainne Matthews and Terry McSween, Quality Safety Edge.

Step 3: Review the trend (line) graph of the checklist practices that result from steps 1 and 2 for the last 12 months. Have these practices been less then 100 percent safe over time or is the last reporting period an anomaly?

Step 4: Compare the observation data and injury reports. Of the practices that were observed to be unsafe, which have caused injuries since the process began and which could cause particularly severe incidents or injuries? Often you will want to select a behavior that is not the most frequent concern because of the potential severity of incidents that might result from that at-risk behavior. Although practices worthy of special attention vary by industry, pay special attention to behaviors associated with fall protection, lockout/tag-out, equipment guards, lifting heavy loads, trenching, and confined-space entry.

Step 5: Select the target practice. Combine the information from steps 1 to 4 to select a behavior to target for improvement based on the probability of injury (based on the number of injuries that have resulted), the severity of potential injury, exposure (the number of times employees perform the practice), consistency (based on the percent safe on a particular concern), and trends in your data.

Step 6A: Gather additional information. Does the steering committee have enough information to complete the ABC analysis? Steering committees often find they do not have enough information to complete a thorough and accurate ABC analysis. In such cases, committees will need to postpone the analysis and gather additional information so that they can complete the analysis at the next meeting.

22.2 METHODS OF GATHERING ADDITIONAL INFORMATION

1. Interview the employees who perform the tasks that require this practice.
2. Observe situations where employees perform this practice.
3. Interview employees who have been involved in incidents.
4. Interview employees who analyzed the incident.
5. Read any policies and procedures, manufacturer's instructions, job safety and job hazard analyses, and other materials that might describe the practice. Talk to experts about the practice.

Step 6B: Conduct an ABC analysis.

(a) Pinpoint the safe practice:
   (i) in general across the department or plant or
   (ii) for the specific situation described in observations and/or injury reports.

If the steering committee decides on (i), identify all the *potential* antecedents and consequences for the practice. If it chooses (ii), identify the *actual* antecedents and consequences for the practice in that situation. The steering committee members use their experience and knowledge of the situation as well as the comments made on the observation checklists and the descriptions on the injury

reports to identify antecedents and consequences. Criteria for a good pinpoint are as follows:
- *Observable*: One could see the action or the result of the action.
- *Specific*: It is enough that two people can agree on it. Not a vague label.
- *Objective*: It is without interpretation or judgment. Not subjective.
- *Active*: Describes the desired actions or results. Does not include any negatives such as *not*. Apply the "dead man rule": If a dead man can do it, it is not behavior.

(b) Pinpoint the opposite, unsafe practice and repeat the analysis.
(c) Analyze the antecedents and consequences that
- support the safe practice,
- suppress the safe practice,
- support the unsafe practice, and
- suppress the unsafe practice.

Ensure that all possible factors that discourage the safe practice and encourage the unsafe practice are covered.

Step 7: Work with management to address facility or procedural issues that contribute to the unsafe act that you are targeting for improvement. If the potential injury is severe or if the exposure is high, then look for ways to redesign the job to eliminate or reduce exposure to the risk or hazard. This may involve other groups, such as engineering and maintenance, and typically goes beyond the scope of the steering committee. Your safety professional will be familiar with the concept of hierarchy of controls that should be considered in addressing risk. The point is that while such interventions are usually beyond the scope of your steering committee, you must consider how to best change the work environment to support safe practices. When the risk can be eliminated or designed out of the job in a cost-effective manner, this is the best solution.

Step 8: Plan the behavioral improvement project:

(a) List all the missing or weak contingency elements (antecedents and consequences) from the behavioral analysis.
(b) Rank them by their influence on the practice.
(c) Identify solutions to the two or three most important missing or weak antecedents or consequences.
(d) Choose two or three solutions and specify actions to recommend to management or for the steering committee to implement. Include solutions that may be difficult, time consuming, or expensive to implement but might be the most important solutions. Also include several solutions that will be easy, quick, and inexpensive to implement even if they will not have the greatest impact. Achieving some early and visible success will encourage steering committee members, participating employees, and those who implement the improvement projects.

(e) Assign steering committee members to implement, arrange for implementation, or recommend implementation to management. If solutions are recommended to management, especially with expensive or time-consuming improvements, prepare documentation of the analysis and the rationale for the choice.

(f) Assign steering committee members to track action items and designate review dates or deadlines. Review improvement project action items regularly to recognize those who are supporting the solution.

22.3 IDENTIFYING WEAK OR MISSING CONTINGENCY ELEMENTS

- **Antecedents**

    Because weak or missing antecedents can often be addressed with a one-time action such as training, job aids, or equipment repair, antecedent solutions are easiest to implement. Table 22.1 lists antecedent approaches.

TABLE 22.1. Common Antecedent Problems and Solutions

Antecedent Problem	Antecedent Solution
Skills and Knowledge	
• Do employees know when and where to use the practice?	Provide education that includes relevant job theory and factors. Consider such job aids as signs, posters, checklists, and other training devices.
• Can they perform the practice?	Provide training that includes practice and demonstration of mastery. Also, ensure the selection process identifies individuals with the physical capability to perform the task safely.
Tools and Equipment	
• Are tools and equipment available?	Provide tools and equipment.
• Are they in good condition?	Replace or repair tools and equipment.
• Are they easily and quickly accessible?	Provide additional access or change location.
Time and Space	
• Do the employees have time to perform the practice safely?	Redesign the work process or procedure.
• Do they have the kind of workspace needed to perform the practice?	Re-engineer the work area or the workflow.

- **Consequences**

    Address weak or missing consequences next because these often require ongoing effort, which can be difficult to sustain. Effective consequences are either certain (frequent) or immediate. Ineffective consequences are uncertain (infrequent) or delayed. The most important feature of consequences is their probability of or how certain they are in following the practice. Size also matters: Large consequences are more effective than small consequences. Table 22.2 lists the consequence approach in order of preference.

    Providing positive consequences will be most effective for improving your process, that is, for supporting quality observations, participation, and goal achievement (see guidelines for recognition and celebrations at the end of this chapter). In fact, consequences for most safety practices are difficult to manipulate. Adding

TABLE 22.2 Common Problems and Solutions for Addressing Consequences.

Consequence Problem	Consequence Solution
1. *Concept question*: Are too many effective negative consequences suppressing the safe practice? *Practical question*: Is performing the job safely uncomfortable, difficult, or slowing down task completion?	*Conceptual solution*: Remove negative consequences for safe practice. *Practical solution*: Redesign the job so that performing the job safely is more comfortable, easier, or quicker.
2. *Concept question*: Are there effective positive consequences for the unsafe practice? *Practical question*: Is the unsafe behavior more comfortable, easier, or quicker?	*Conceptual solution*: Remove positive consequences for unsafe practice. *Practical solution*: Redesign the job so that performing the job unsafely is less comfortable, requires more effort, or takes longer.
3. *Concept question*: Are there enough effective positive consequences supporting the safe practice? *Practical question*: Are positive consequences designed into the task? Do observers and supervisors consistently provide positive feedback for the safe practices?	*Conceptual solution*: Add frequent (certain) positive consequences. *Practical solution*: Redesign the job so that positive consequences are built into the task. Ensure observers and supervisors are monitoring and providing positive feedback on the safe practice.
4. *Concept question*: Are negative consequences for the unsafe practice ineffective? *Practical question*: Are negative consequences designed into the task? Is coaching certain? Do supervisors understand and consistently enforce disciplinary policies when appropriate?	*Conceptual solution*: (*Last resort.*) Add effective negative consequences. *Practical solution*: Redesign the job so that negative consequences are built into the task. Ensure observers and supervisors are monitoring and providing corrective feedback on the safe practice. Ensure disciplinary action is taken when appropriate.

tangible consequences that are soon, certain, and positive for performing a task safely is usually neither desirable nor practical. If you are unable to redesign the job so that it provides better built-in or natural consequences (such as when you find eye or hearing protection that is more comfortable), then you will often need to try to increase the frequency of feedback on the targeted behavior through your observation process and site leadership. As you track your success toward achieving your improvement targets, you will need to share data on progress with the employees affecting that improvement, celebrate goal achievement, and recognize individuals who contributed or assisted with implementation of your action plan.

Step 9: Use the process to support the intervention:

1. Meet with employees to explain why the steering committee chose the new target practice (show the data) and what the committee plans to do or to recommend be done (show ABC analysis). Ask employees if they have any other suggestions.
2. Guide employees in setting a reasonable improvement goal for percent safe on that practice for the next reporting period (see guidelines for setting goals below).
3. Ask employees to choose a celebration for achieving the goal from a list of two or three options.
4. Ask observers to focus on the new target practice with observation and feedback.
5. Post the trend (line) graph of the target practice with the goal marked for the next reporting period. Update the graph as soon as the reporting period is complete. Discuss the data during department or plant meetings.
6. Schedule the celebration as soon as the goal is met. Recognize individuals for working toward the goal even before it is met and especially during the celebration.
7. Recognize individuals who implement recommended interventions (e.g., maintenance staff, engineers, trainers, and managers).

22.4 GUIDELINES FOR SETTING GOALS

Review all safety data at regular steering committee meetings to identify checklist practices to be targeted by improvement projects or process measures to be addressed, such as percentage of participation.

Lead employees in setting goals regarding a targeted practice or other measures during area meetings:

1. Solicit input also from employees who are unable to attend area meetings.
2. The smaller the group to which a goal applies, the more effective it is in motivating the group's behavior. The larger the group, the less impact each individual feels his or her behavior will have on achieving the goal. The best practice is therefore to have each area set its own goal. (However, this

practice requires separate graphs and data analysis for each area and can be time consuming.)

3. If goals are to be set for the entire organization, the steering committee will need to obtain input from all employees, which can be difficult. The best approach to is hold one large meeting where everyone sees the data and decides the goal. (However, it is usually necessary to obtain the input of each shift, department, or team separately and combine them in some way—such as by using the average suggested goal—to set the organization goal.)

Ask employees to select goals by determining current performance in relation to the baseline on the trend (line) graph, where they would eventually like to be (the ultimate goal is usually 95 to 100 percent), and reasonable steps to get there:

1. The first goal is best set at the maximum level obtained during baseline.
2. Subsequent goals are set in increments no greater than the current variation from a prior goal. For example, if the maximum variation in percentage of Participation were 5 percent, then a reasonable subsequent goal would be 5 percent over the current level.
3. New goals are set only when current goals are met. Thus, current goals may be in effect for varying lengths of time.

Once the next goal is chosen, ask employees to select a celebration from several choices upon attaining that goal. If a novel celebration is suggested (i.e., one not on the list of options), the steering committee may offer to explore that as a possibility for the next goal.

Mark the chosen goal and celebration on graphs of the percent safe for that target practice. Provide feedback on progress toward the goal by updating graphs and discussing data in meetings.

22.5 GUIDELINES FOR RECOGNITION AND CELEBRATIONS

Another steering committee primary responsibility is ensuring that behavioral safety remains a positive process. This means that employees participate because they want to, not because they have to. Positive reinforcement, primarily in the form of individual recognition and team celebrations, will be the primary tool to ensure a positive process. In Chapter 11 we talked about designing an effective recognition and celebration process. The following guidelines will help the steering committee in administering that process:

- Effective recognition and celebrations are based on achieving a goal or meeting a criterion, not on the passage of time. Once-a-month parties, employee-of-the-month selections, and periods of time without an injury are examples of time- instead of performance-contingent recognition and celebration programs.

- Recognition and celebration may occur frequently or infrequently depending on whether the established goal or criterion has been met.
- Recognition is earned by performing desired behaviors to meet a specified criterion—completing 10 observations, for example. Celebrations are earned by achieving specified goals—60 percent participation, for example. Both are determined objectively, not judged subjectively.
- Recognition and celebrations are win–win propositions. Everyone can earn recognition by meeting the specified criterion and everyone in the group can celebrate when a goal is achieved. Recognition is not awarded only to the employee or group that was first or best but to anyone who meets the criterion and any group that achieves the goal. Whether an employee earns the recognition or a group earns the celebration is independent of whether another employee or group earns it.
- Recognitions and celebrations may be unannounced or "planned spontaneous;" that is, the steering committee decides in advance what will be recognized during a specified reporting period and looks for opportunities to reinforce such individual performances as conducting an observation, completing a quality observation, doing more than scheduled observations, observing a specific department or task, and the like.
- Both individual recognition and organization celebrations can be formal or informal. Formal recognition might include placing commendatory letters in an employee's personnel file; making positive comments on a performance appraisal; increased compensation by means of a bonus or raise; making choice assignments to locations, projects, or departments; or even promotion. Informal recognition might include individual thanks and commendation in private, public acknowledgment, a letter to the employee's family, and posters on bulletin boards that recognize an employee or employees. Formal celebrations might be company dinners while informal celebrations might be free barbecue in the lunchroom.
- Social or personal (i.e., nonmaterial) recognition is best for avoiding common employee concerns:
  1. An awarded material item being less valuable than awards given to others.
  2. The value of an awarded item not being equal to the value of the performance.
  3. Some employees receiving more recognition than they deserve while others receive less.
  4. Some employees performing the desired behavior only to receive material items.
  5. An awarded item not being meaningful to the individual.
- If symbols or tokens of recognition are used, choose those that are free or almost so. Money is best saved for celebrations. Tokens do have one advantage: They can facilitate interaction between the employee giving and the employee receiving the recognition. Such interaction is often especially

important early in the behavioral safety process when giving and receiving recognition may be new to many employees.
- Celebrations are most effective when:
  1. They bring everyone together in one place at one time for food and refreshments.
  2. They pinpoint the reason for the celebration.
  3. They include a discussion of the goal and how it was met followed by setting the next goal.
  4. They recognize individuals during the celebration who contributed to meeting the goal.
  5. They involve significant people, such as respected executives who make the awards, well-liked employees who tell the story of their involvement, or family members who describe the value of the safe practice.
  6. Everyone is invited to a celebration, not merely those who contributed to achieving the goal. Doing so helps encourage everyone to participate in achieving the next goal.
- Customize recognition to the employee receiving it:
  1. Would this employee prefer to receive recognition from a steering committee member, teammate or colleague, supervisor or manager, organization manager, safety professional, or someone else completely different?
  2. Would this employee prefer private or public recognition?
  3. What small token might signify recognition and remind this employee (and others) of the recognized behavior?

23 Other Support Programs

The list below identifies other process components and programs that may be useful in supporting your safety improvement efforts. The process components are important, and each is worth considering as a key element of your safety process. The other programs represent a sample of optional activities that may be part of your initial promotional efforts when kicking off behavioral safety or they may be useful elements of action plans that the steering committee develops to assist in achieving its improvement targets.

23.1 ADDITIONAL SAFETY PROCESS COMPONENTS

- Process safety reviews
- Near-miss programs
- Involvement in incident investigations
- Safety suggestion systems
- Safety orientations for new employees and contractors

23.2 SUPPLEMENTAL SAFETY PROGRAMS

- Contests to name the behavioral safety process
- Worker-designed safety slogans and logos
- Small group discussions of values
- Presentations from process champions
- Safety process conference
- Safety day celebrations
- Safe-behavior promise cards
- Public safety declarations
- Verbal feedback support cards
- Tokens for awards or lottery
- "Safety share" discussions

Note: The author is indebted to Scott Geller (1989) for many of the ideas presented in this chapter. Scott is one of the true gurus of the behavioral safety field.

- Off-the-job safety programs
- Presentations or question-and-answer sessions with management representatives
- Creation of a behavioral safety process support group

Process safety reviews, involvement in incident investigations, and the safety suggestion system are key elements of the safety process for many organizations. You may wish to consider several of the suggested programs as part of your initial implementation in helping to reduce resistance to your new behavioral safety process. For example, you might decide to have several small group discussions on values prior to your initial kickoff meeting.

In addition, you can arrange a variety of education and training programs—for example, as for reviewing standard safety practices both on and off the job. Examples of on-the-job educational programs include courses on lifting and back safety, protective equipment, first aid, and emergency response. Off-the-job safety programs might include courses on yard care safety, common household incidents, gun safety, boating safety, and defensive driving.

23.3 ADDITIONAL SAFETY PROCESS COMPONENTS

Process Safety Reviews. The real key to safety is designing it into the process. Safety has to be engineered into the design and layout of the site. Use process safety reviews to ensure fail-safe engineering design of equipment and controls. Such reviews are technically complex studies that involve careful analysis of "what if" questions about what could go wrong at each step of a process. They need to be a standard step for all new construction and debottlenecking efforts. Depending on the nature of your industry, you may also need a group that conducts a regular process safety review in each area.

Near-Miss Program. A near-miss program is an important component of a total safety process. Steering committees should investigate near-miss incidents with the objective of preventing actual future occurrences. The key to a successful near-miss program is establishing a positive system that encourages honest reporting. Near misses can be the basis for case study discussions in safety meetings. Employees might develop video tapes showing near-miss situations for use in training and for discussion in safety meetings. In developing case studies or video tapes, be careful not to make the process punishing to employees who have honestly reported such events.

Involvement in Incident Investigations. Representative employees should participate in investigating incidents in their work areas. They are often in the best position to help identify and understand the factors that led to the incident. In addition, their involvement helps create ownership and acceptance of the investigation report and resulting recommendations. Their participation in the investigation positions them to help explain what happened so that others in the area can more fully appreciate the report and the rationale for new procedures and preventive measures.

Employee participation in incident investigations can further result in more accurate conclusions from such investigations. In a recent project involving a bagging operation, several members of a behavioral safety process design team participated with the incident review board to investigate an incident that occurred during the final design and implementation of a new safety process. The design team members were employees who worked in the area where the incident occurred. They were able to verify that the procedure that the employee had followed was common practice even though not part of the documented procedure. As a result, the incident investigation faulted the system and recommended that supervisors and employees in the area change their work practices to match the formal procedure. Without the input of these employees, however, the review board would have accepted a preliminary finding that the employee was at fault for not following the correct procedure.

Safety Suggestion Systems. For a behavioral safety process to gain credibility, management must be proactive in addressing safety-related maintenance issues. Employees must see such issues being systematically addressed or at least see management's plan for addressing them over a period of time. A safety suggestion system is one way to help get employees involved in identifying such issues.

Employees are often in the best position to see and identify safety hazards in their work areas. A safety suggestion system is a good way to identify such hazards, particularly maintenance items and hazardous conditions that need correction. Such a system requires a process for reviewing suggestions, assigning a priority to those that need to be addressed, tracking their status to completion, and providing feedback to the employee(s) who originated them. The steering committees also should provide recognition to employees who make significant safety-related suggestions.

Various studies have documented that the key to making suggestion systems work is providing a clear, immediate response to each suggestion and keeping employees updated on the status of their suggestions. Posting a summary report that lists all active suggestions and their status is often the best way to keep employees so informed. Generally, someone on the safety team takes responsibility for tracking the status and resolution of employee suggestions, posting an updated summary for employees, and reporting the status to either the steering committee or management.

Safety Orientations for New Employees and Contractors. The orientation of new employees is a critical element of a safety program, and too often safety gets lost in an orientation that is filled with dealing with a variety of human resource issues ranging from company policy to insurance forms. Employees should receive a checklist that includes a list of specific safety items that they need to discuss with their supervisor when they enter the work area for the first time. This is an area to which many organizations need to pay more attention. The observation process is an ideal method for providing safety orientations for both new employees and contractors. New employees can review the observation checklist with their supervisors and participate in the observation process as part of their own orientation. The observation checklist can also be a formal part of the start-of-work meeting with new contractors or contractor employees. It can help clarify the expectations for contract help and lead into a discussion of the observation process and how you will be providing additional follow-up on safety compliance.

23.4 COMMON SUPPORT PROGRAMS

Contest to Name Behavioral Safety Process. While designing the observation process or shortly after implementation, consider holding a contest to name your behavioral safety process. Such a contest is usually coordinated by the design team, which will have to plan the rules for the contest and the procedure for submitting a name, campaign for promoting the contest to employees, choose the prizes for winners, and determine how entries will be judged. Such a contest is a good way to begin introducing the behavioral safety process to employees.

Worker-Designed Safety Slogans and Logo. In another common safety program the steering committee holds a contest for employee-generated safety slogans and/or logos for your behavioral safety process. The winning slogans and logos can become the basis for signs, electronic message boards, T-shirts, or posters promoting the behavioral safety process within the plant or work area.

Small Group Discussions of Values. After your design team has completed its work on values, you will need a process for communicating the new values to all employees. One of the most effective ways to accomplish this objective is to involve small groups of employees in discussions about practices that support each value statement. Small group discussions can be conducted during normal safety meetings with a spokesperson for each group presenting a summary to the other groups. Before the discussion begins, ask employees to ensure that their practices are concrete and observable (see the discussion on pinpointing in Chapter 21). Consider conducting this type of small group discussion exercise prior to your kickoff meetings.

Presentations from Process Champions. During the kickoff meeting or special safety meetings after this meeting, you may want to invite individuals from other parts of the company to discuss their experiences with the behavioral safety process. These guests could be members of the design team, representatives from management, or employees who participated in successful implementation efforts elsewhere in the company. (Note that giving employees an opportunity to visit other locations to talk about an area's successes is a good way to provide recognition for some of that area's participants.) By discussing the similarities and differences between how each site implemented the process, you can learn how to enhance your own efforts and possibly avoid some mistakes made by others.

Safety Process Conference. Your company might hold a companywide safety conference. The conference agenda might include a combination of training sessions, a celebration, and social events. The format might include several blocks of concurrent training sessions, presentations or posters from safety teams describing their successful efforts, and a catered luncheon or picnic with an entertaining keynote speaker.

Safety Day Celebrations. Companies will often sponsor a day-long celebration to kick off special programs or applaud the success of their efforts. Activities can include a company picnic with games and guest speakers, a conference with meetings and seminars, or some combination of events such as both of these along with dinners, mixers, and award ceremonies.

Safe-Behavior Promise Cards. Safe-behavior promise cards can be an effective way to initiate a program. They may identify a specific behavior, such as proper lifting or correctly using seat belts, or are complex sets of behavior such as compliance with a list of pinpointed safe practices. The idea is to get employees to make a personal commitment to a specific goal, ideally for a fixed period. For example, you might decide to have a seat belt campaign during April and ask everyone to commit to wear their seat belts 100 percent during the month.

Public Safety Declarations. This variation on the safe-behavior cards can take several forms. Again, the purpose is to have employees make a public commitment to achieving a specific compliance goal. It might be in combination with a safety day ceremony on a stage during which employees shake their manager's hand and make a public commitment to compliance with safety procedures and perhaps sign a written statement to the same effect that is later posted on the area's safety bulletin board.

Feedback Support Cards. Written cards can be an effective way to support verbal feedback as part of the observation process. They may be thank-you cards distributed to employees who are performing safely on the day they are observed. Observers may give the cards to employees to support a specific safe behavior that is being emphasized by a special campaign, such a proper lifting in conjunction with a back safety campaign. Or the observers may give the cards to everyone in an area with 100 percent safety performance.

Such thank-you cards can also be part of the recognition program. They may be redeemable for a drink in the cafeteria or might be part of a lottery as described below. Another approach is for the company to contribute money to charity based on the number of cards earned by employees. For example, for each thank-you card earned by employees, the company might drop 25 cents into a collection bowl. Hard-hat stickers on one side of the thank-you cards might be used to provide

COMMON SUPPORT PROGRAMS 213

THANK YOU FOR WORKING SAFELY

You were observed working safely. By working safely you are being recognized by management for following company procedures and policies and for protecting yourself, fellow workers, and family.

The company would like to recognize your actions, so please present this card to your foreman or safety department.

Figure 23.1. Sample positive-feeback card.

recognition to employees that are contributing to this process. Figure 23.1 presents an example of a positive-feedback card that might be part of your behavioral safety effort.

Similar cards can also facilitate corrective feedback. Cards for this purpose may take several forms. Some companies provide a number of corrective feedback cards to employees to use independently of the observation process, with the idea that they pass them on to other employees who they see are working unsafely. Other cards make a personal statement, such as "Asking me to look the other way is like asking me not care, and you are much too important" or "Asking me to overlook a simple safety violation would be like asking me to compromise my entire attitude toward the value of your life."

Du Pont uses STOP cards as part of its safety process. Employees complete and turn in these cards to document the types of unsafe practices that they observe but do not include the name of the person or persons performing the unsafe activity. The problem, as mentioned earlier, is that these cards focus almost exclusively on what employees are doing unsafely, rather than what they are doing right.

Tokens for Awards or Lottery. This support device is similar to the feedback cards except that the employees receive tokens that may entitle them to prizes or special awards. The tokens may be, for example, trading stamps or lottery tickets. They may have a specific value, such as stamps or points that are used to earn specific awards, or they may allow those who earn them to participate in an award drawing.

Safety bingo is one common application. As emphasized throughout this book, such efforts are more effective if they are based on safety compliance rather than simply on going without an incident. These approaches can obviously be combined, so that bingo tokens are earned based on observations but the game ends if anyone has an injury.

"Safety Share" Discussions. Safety-sharing discussions can provide an effective safety meeting agenda. After the meeting gets started, simply go around the room and have all participants relate something safety related they have done within the last 24 hours (or 48 hours or even the past week).

Your steering committee might plan a take-off on such discussions around specific aspects of the observation process. For example, observers might recognize groups of employees who achieved 100 percent safety or specific individuals for particular safety practices. In addition, observers may want to share specific unsafe activities or conditions that they believe other employees should be aware of. As with other aspects of the behavioral safety process, you should take care not to identify specific individuals who were observed engaging in an unsafe practice. These discussions should be kept positive and educational, always avoiding embarrassment or humiliation.

Off-the-Job Safety Programs. Not all of your efforts need to focus on safety in the workplace. You can also create programs focused on safety off the job. These programs might simply consist of a safety meeting in the spring on boat safety or at the beginning of the hunting season on gun safety. Or they may be more complete campaigns targeting seat belt use or preventing off-the-job back injuries.

For example, some plants have initiated seat belt campaigns with posters, safety meeting training films, and other activities. In some cases, the site management staff has arranged to stand at the entry gate and pass out silver dollars or other tokens to everyone entering (or leaving) the parking lot wearing their seat belt. [For additional suggestions on increasing seat belt use, see Geller (1984), Geller and Hahn (1984), and Geller and Lehman (1991).]

Presentations or Question-and-Answer Sessions with Management Representatives. A common complaint from employees is lack of direct contact with management. One way to enhance such interactions and support your behavioral safety process is to arrange from time to time for a member of upper management to lead one of your regularly scheduled safety meetings. You might ask these managers to prepare a brief discussion on their vision or expectations regarding safety and to provide time for questions and answers. Your steering committee might develop a list of questions, then arrange for your manager to prepare for and speak to the questions on the list in case the employees do not spontaneously generate many questions.

Creation of a Behavioral Safety Process Support Group. You may wish to create a behavioral safety support group within your organization. This group could include representatives from several locations who would serve as a sounding board and resource for one another. On a larger scale, you could provide a forum for several groups to share their successes and discuss how they have handled issues such as the implementation process and its challenges.

23.5 COORDINATE SPECIAL PROGRAMS

The key to maximizing the value of your safety efforts is to coordinate your ongoing safety process and your special programs. Thus, if your plant typically experiences a high rate of incidents during the summer, you might have a special program for the "dog days of summer," for example. You might design a kickoff meeting with a presentation from management, use a safety pledge card for an appropriate time period, conduct a slogan contest related to the dog-days theme, and give every employee one thank-you card, asking them to hand the card to another employee during the time period, noting on the card why it was given. At the end of the period you might then arrange a special celebration if the area proved successful in not having an incident and provide awards to the employees earning the most thank-you cards.

24 Long-Term Case Studies

When the first edition of this book was written in 1993, data were not available on the long-term success of the behavioral approach. The two case studies that follow show that the process can be long lasting and help support a sustained level of safety performance for long periods of time.

24.1 BEHAVIORAL SAFETY IN A REFINERY[1]

Implementing a value-based safety process (VBSP) in a large organization with many self-contained areas presents numerous challenges, but the long-term success of this behavioral safety process speaks for itself. A major oil refinery with two large facilities in the United States overcame these obstacles in an effort to improve safety performance. Before behavioral consultation was sought, both plants relied primarily on traditional safety methods, such as focusing on lost-time incident rates, to prevent safety incidents. Despite these safety efforts, there was ample room to improve safety practices.

The initial plant is comprised of six self-contained areas: (1) power thermal, (2) central maintenance, (3) catalytic cracking unit (CAT), (4) acid-alkylation treatment unit (AAT), (5) oil movement, and (6) reformer. When the VBSP was initially implemented, this facility had approximately 1700 employees.

Phase 1: Pilot Area

The power thermal area of the initial plant, with approximately 200 employees, was chosen as the pilot group because it held the most room for improvement. In 1992 alone, this area had five recordable injuries and three lost-time incidents.

The steering committee for the power thermal area consisted of 10 hourly wage employees and 1 manager, and the committee leader was always an hourly wage employee. The committee held weekly, or sometimes biweekly, meetings. Consultants trained the steering committee members in VBSP methodology in three days, then together designed the safety process for each unit within the area in sessions of two days a week over three weeks, a total of six days. The VBSP training of all employees, from the managerial level down, was accomplished by

---

[1] This section was written by Wanda Myers and Terry McSween, of Quality Safety Edge, and Alicia Alevero, of Western Michigan University.

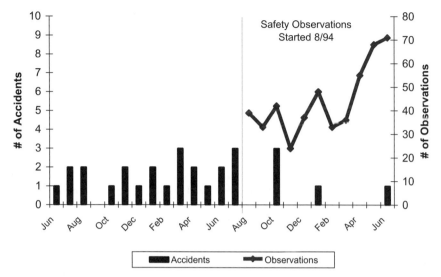

Figure 24.1. Inverse relationship between number of safety observations and number of accidents during the initial implementation.

training 20 employees at a time over 8-hour periods. Conducting safety observations was strongly emphasized, and all employees were required to practice conducting observations using videos of their own work areas. After training, participation as a safety observer was strictly voluntary.

The steering committee recorded three primary measurements from the start of implementation: (1) the number of observations conducted, (2) the percentage of participation (the latter defined as conducting two observations a month), and (3) types of safety concerns. Figures 24.1 and 24.2 clearly depict the success of the

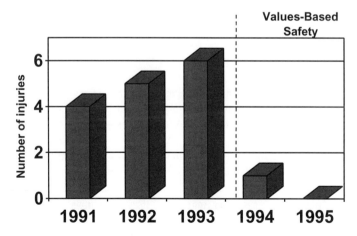

Figure 24.2. Recordable incident rates for the thermal area before and after VBSP implementation.

VBSP implementation. Figure 24.1 demonstrates an inverse relationship between the number of safety observations conducted and the occurrence of accidents. The dramatic decrease in the number of recordable incidents is shown in Figure 24.2. In fact, the single recordable injury in 1994 occurred during the first quarter of the year, prior to implementing the behavioral safety process. The power thermal area had no recordable injuries for two years after implementation and no lost-time incidents four years after implementation in the pilot area.

Phase 2: Plantwide Implementation

The VBSP helped the site achieve and maintain a low rate of injuries (see Figure 24.3). The five remaining areas followed the same approach outlined above. Central maintenance began implementation in 1994, CAT followed in 1996, AAT commenced in 1997, and the oil movement and reformer areas began in 1998. To overcome the communication barriers that often result from having many self-contained areas, the steering committees designed and utilized several techniques to encourage critical communication. All areas have bulletin boards within each unit solely dedicated to the VBSP. Graphs of the three primary measurements, campaign or contest announcements, and minutes of steering committee meetings are frequently updated and posted within each area. In 2000, the plantwide steering committee began to distribute a newsletter and developed a safety website to further ensure dissemination of the organization's safety progress. Management at the initial plant also provides funds specifically for the continuous improvement (recognitions, celebrations, etc.) of the values-based safety process, another key element in the long-term success of its safety process. The benefits of hard work and dedication toward the successful implementation of a behavioral safety process

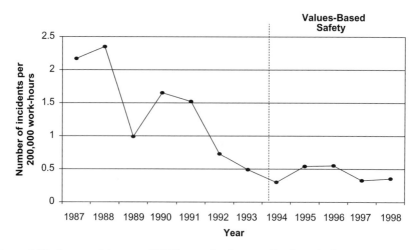

Figure 24.3. Lost-workday case (LWC) rates for the overall refinery before and after VBSP implementation.

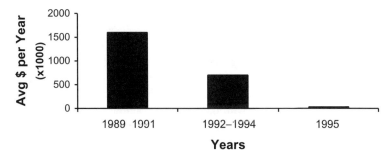

Figure 24.4. Direct costs of injuries for the years before and first year after implementing behavioral safety.

are countless, but the financial benefits are one of the most dramatic. Figure 24.4 shows the dramatic decreases in worker's compensation costs after the successful implementation of VBSP.

Because of our success at the original site, we were asked in 1998 to also upgrade the safety program at a second facility. At that time it had approximately 800 employees and four self-contained areas: a terminal, a west plant, and two east plants. The terminal area was already using behavior-based safety, but the process was floundering because of infrequent observations, poor recordkeeping, and insufficient communication. For this reason, we began VBSP implementation with the west plant. The process used was similar to the one described throughout this book with modifications, such as the ones described earlier, made to overcome the challenges of an organization with numerous self-contained areas. Terminal area employees attended the observer training provided for the west plant employees and worked to revive their behavior-based safety process. In 2001, the VBSP was also implemented in the two east plants.

Recently, the corporation, owner of approximately seven businesses worldwide, identified behavior-based safety as a best practice and encouraged its implementation throughout the organization. This was a convincing validation of the success of the VBSP described in this book.

24.2 EMPLOYEE SAFETY PROCESS AT AN ORE-PROCESSING FACILITY[2]

The production of alumina, which is extracted from bauxite ore, is a business with many hazards. Employees work in situations where they are at risk of burns from caustic chemicals, hot liquids, and extremely high heat as well as of injuries resulting from the traditional hazards common to most industrial settings, such as slips and trips. This particular plant, however, had more problems than most. It was

---

[2] This section was written by Judith E. Stowe and Terry McSween, of Quality Safety Edge, and Angelica Grindle, of Western Michigan University.

built in the early 1940s, and the company had not invested adequately to keep the facility in good mechanical condition. As a result, the plant was in such bad physical condition that workers had to wear acid suits just to walk into the production areas. The organization had cut its injury rates in half the year before we began the project but still experienced a significant number of lost-time injuries. Both the management and the safety committee were very interested in achieving further improvements.

Given the serious problems with the physical condition of the plant, we were a bit hesitant to undertake this project. However, employees believed that management was serious about improving safety, and we became convinced that management was serious about safety and would do all it could to maintain and upgrade the facility in conjunction with the behavioral process. These management efforts helped the behavioral process to be successful.

To assist in improving the safety process, we first conducted a safety assessment as detailed in this book. A preestablished company safety team became the design team. Most of its 15 members were hourly workers. Quality Safety Edge consultants conducted a three-day workshop on values-based safety and then facilitated four two-day meetings to design a safety process that included the following components:

- **Training**

One goal of the safety process was to have all employees trained in conducting observations. However, given the large number of employees, a novel approach to observation training had to be taken. To accomplish our goal, we held train-the-trainer sessions in which we trained 12 area safety representatives on how to conduct observations, manage safety data, create action plans, and deliver feedback. These safety representatives then trained the other employees in their areas.

- **Kickoff Event**

Following safety representative training, the company scheduled a huge weekend family day to kick off the new safety process. Employees and their families were invited to a park the company had reserved for the event. The plant manager participated and safety representatives were introduced. The following Monday, three large site meetings were held to explain the process and trained safety representatives conducted small area meetings to discuss questions and details.

- **Observations**

Multiple levels of the organization were involved in the observation process. The goal was to have hourly employees and general forepersons conduct at least one observation per week and managers and department heads conduct at least three observations per week. Conducting observations and being observed were voluntary.

- **Feedback**

Observers gave immediate feedback to the employees observed. Percentage data on concerns and safe acts were then posted on the area safety boards. The crew safety representatives would then review the data at monthly safety meetings and send the accumulated observation sheets to the process steering team for review every three months. In addition, safety performance and safety issues were addressed during regular toolbox meetings.

- **Recognition**

Area team celebrations were held based on safety process milestones. The process steering team monitored area team progress and established criteria for recognition. Awards ranged from Starbucks gift certificates to paid time off work. As the process has matured, recognition items have changed to such things as T-shirts, coolers, and jackets for participation combined with no lost-time accidents.

- **Safety Communication**

The company regularly distributed a newsletter that included items and anecdotes regarding the safety process, safety performance levels, and anonymous discussions about the causes and prevention of any injuries that had occurred. The newsletter also provided recognition items and information about improvements initiated as a result of the behavioral process and stressed the importance of safe work practices at all times. Each group's overall safety performance was discussed and safety improvement strategies were developed during monthly steering team meetings and shared with work crews in their monthly safety meetings. A steering team member usually also attended each of these monthly departmental meetings.

Although it took several years to train the majority of employees to conduct observations, over 10,000 observations are currently conducted each year. The graph in Figure 24.5 presents data in the initial phase of implementation. It also shows the critical inverse relationship between the number of observations and the number of injuries. That is, as the number of observations increased, the number of

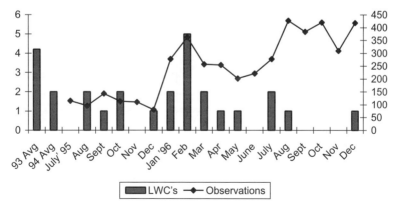

Figure 24.5. Data showing the inverse relationship between observations and lost-workday cases (LWCs) during implementation at an ore-processing plant.

222 LONG-TERM CASE STUDIES

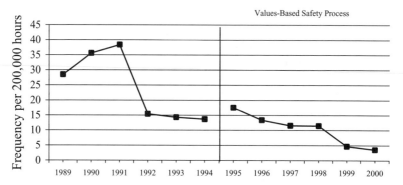

Figure 24.6. Recordable injury rates before and after implementation of a behavioral safety process (OSHA incident rate).

accidents decreased. These data helped management and employees build support for greater participation and an increased frequency of observations.

As shown in Figure 24.6, since its implementation in 1995, the safety process has demonstrated continued success for more than six years. Moreover, in the last few years the company has twice achieved over one million hours without a lost-time accident.

The company not only has achieved significant reductions in both OSHA-recordable and lost-time incidents but also has had a significant reduction in accident severity. The graph in Figure 24.7 reflects a steady decline in severity and fairly stable results for the last few years.

Another important lesson learned from this project led to this being the last time we used a mentoring approach to train observers. We typically achieve much higher levels of participation when employees participate in formal training led by consultants or internal trainers. Formal workshops not only train employees in how to conduct observations but also impart why observations are so important. Although the project was ultimately successful, building support and generating participation in observations took much longer than is typical of our implementation efforts.

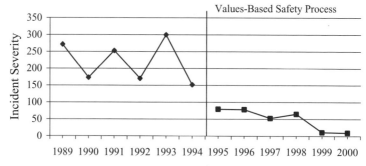

Figure 24.7. Severity rate of injuries before and after implementation of the behavioral safety process.

25 Self-Observation Case Studies[1]

An area of behavioral safety that has only recently begun to get attention from researchers is the application of behavioral safety to employees that work in isolation. One research study has documented that a self-observation-based approach resulted in a significant increase in safe driving practices by bus drivers (Olson and Austin, 2002). The three case studies that follow show that this approach can significantly improve both safe behaviors and safety outcome measures.

25.1 CANADIAN GAS PRODUCTION AND PIPELINE COMPANY

Successful safety implementations are often challenging when dealing with lone workers, but sometimes a simple solution and frequent reinforcement are all that is necessary to achieve dramatic results. Such was the case with an organization that collects and processes natural gas. The behavioral safety process we designed for this organization improved overall safety, interaction between employees and management, and employee participation that additionally brought the employees a sense of empowerment.

At the time of consultation, the organization employed 60 workers who worked on wellheads and gas pipes in isolated areas. Safety statistics for the workers themselves were very good before implementation, most accidents being vehicle related. We therefore designed our implementation to concentrate on two areas: decreasing the number of vehicle accidents and demonstrating the value of the safety process.

Together with a group of volunteer employees we developed a *driving safety index* in $3\frac{1}{2}$ hours. The index listed eight items related to safe driving—for example, wearing seat belts, not exceeding speed limits, and keeping windows clear. Next to each item were three columns for checking yes, no, and not applicable (N/A). A safety percentage was calculated by dividing the total number of boxes marked yes by the total number of marked boxes and multiplying by 100 percent. Drivers were contacted via radio to inform them when it was time to conduct a self-observation. Upon receiving the call, drivers pulled off the road and immediately completed the driving safety index. Observations were conducted three times a week during the first two months after implementation and two times each week thereafter. As often

---

[1] The case studies in this chapter were written by Ann Pinney of Quality Safety Edge and Alicia Alevero of Western Michigan University.

is the case with self-monitoring, some members of the organization were concerned that inflated scores would be reported. We did not believe this a matter of critical concern because, at the least, the radio call to conduct an observation would prompt the drivers to think about their driving practices once more than they would have otherwise.

Goals and reinforcement were managed by behavioral safety representatives (safety reps), a group of volunteer employees, but were not announced to the other linemen, who were simply told they would be rewarded for increases in safety. The original team safety goal, set by the safety reps, was 85 percent. When teams achieved this goal, they were given a reward, such as a sausage-and-biscuit breakfast. Deliveries of rewards were not announced in advance and thus served as "surprise" reinforcers, a tactic that encouraged employees to achieve increases in safety as often as possible in order not to miss the opportunity to receive a reinforcer. Although the deliveries were always a surprise, they did occur frequently enough to maintain high levels of safety performance.

Figures 25.1 and 25.2 show the solid effectiveness of the driving safety index. Figure 25.1 shows the increased average team levels of safe driving performance while Figure 25.2 shows the decrease in vehicle accidents, medical aids, and lost-workday incidents after the behavioral safety intervention. An index was also developed to monitor supervisor behavior. The *supervisor index* listed items related to their role in the safety process, examples being held safety meeting, provided feedback to employees on safety behaviors, posted data on graph, and the like. Supervisors completed the index each time a driver was radioed to conduct a self-observation.

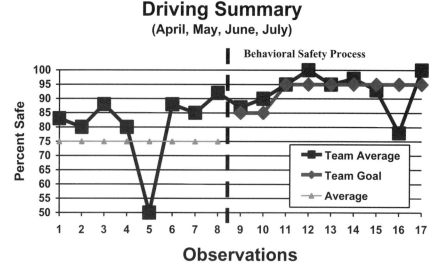

Figure 25.1. Data on the percent of safe driving practices reported during self-observations by employees working in gas production fields and plant.

CANADIAN GAS PRODUCTION AND PIPELINE COMPANY 225

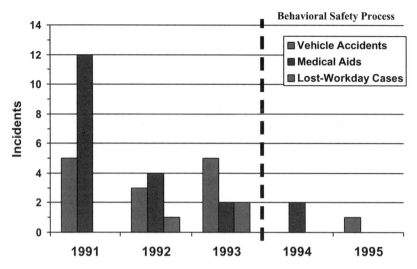

Figure 25.2. Implementing a self-observation behavioral safety process dramatically cut the rates of vehicular incidents, medical treatment cases, and lost-workday cases for employees at a gas plant.

Because a formal positive reinforcement process was already in place, we were asked to tie that existing process to our new safety process. As a result, in addition to the reinforcer for increases in driving safety, three other reinforcers were set in place: peer-to-peer reinforcement, supervisor-to-employee reinforcement, and employee-to-supervisor reinforcement. In order to monitor these forms of reinforcement, each was added to the appropriate index. The driving safety index therefore also included a row to monitor the receipt of positive reinforcement from peers as well as from supervisors (marked off as yes or no). Similarly, the supervisor index included a line to monitor the receipt of positive reinforcement from employees, although unlike the driving safety index, the supervisor index also included a line for supervisors to monitor their own delivery of recognition to employees. Figure 25.3 shows the increased use of positive reinforcement.

The peer-to-peer reinforcement process also provided an example of employee-delivered reinforcement tied to a tangible personalized reinforcer. Employees were given five special certificates per quarter, each worth five "organizational dollars." The certificates were to be given by one employee to another when the former observed a safety-related task that was thought merited recognition. Receivers could use them toward the purchase of any article they wanted by simply attaching the certificate to the purchase receipt for reimbursement by the organization. Attached to each certificate was a tear-off portion that required the observer to list the specific observed action that resulted in awarding the certificate. These tear-off slips were collected, compiled into a list, and publicly posted throughout the facility. Though no names were posted, these lists served the purpose of informing everyone about types of behaviors that resulted in reinforcement. Employees who

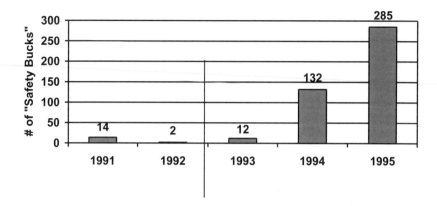

Figure 25.3. Behavioral safety process helped the organization increase the use of positive reinforcement as represented by the number of "safety bucks" delivered by both peers and supervisors.

had not been given certificates used these lists as a valuable information resource. All employees became very judicious about how they awarded their certificates, feeling they needed to be earned instead of merely being given away.

The supervisor-to-employee reinforcement involved deliveries of feedback, praise, and recognition for safety-related accomplishments. Employee-to-supervisor reinforcement also involved deliveries of feedback and recognition as well as gratitude for involvement.

Scorings of calculations and arrangements of surprise reinforcers were done by employee volunteers. All employees provided input regarding their reinforcer preferences, but what was delivered upon goal achievement was a surprise. Employees also provided input concerning the reinforcement budget as to the monies allocated for the purchase of tangible reinforcers. During a safety meeting, 59 of the 60 employees voted to apply the allocated funds to safety reinforcers instead of traditional safety awards (e.g., employees had previously received prizes such as a television after working six months without an injury).

The results of this self-observation process combined with extensive reinforcement were overwhelming. Simply stated, the safety process transformed the organization. There was a dramatic increase in safety awareness, especially among new employees. As a result of employee involvement in developing the driving safety index and managing the reinforcement budget, employees felt an increased sense of empowerment and continued their participation in the process. Evidence of a strong desire to be involved was apparent during the first meeting after implementation. One employee related a concern about being unable to participate because "sometimes I don't get a radio call to do an observation because of poor reception." The aversive relationship with management previously reported by employees was no longer existent. The safety process substantially increased communication between all members and across all levels of the organization.

One employee put it best: "Not only did safety improve, but every other aspect of our business improved."

25.2 ELECTRIC UTILITY

Organizations with unique occupations and a culture that lacks employee involvement require clever and creative behavioral interventions in order to succeed. Such was the case with a large power and light company that employs numerous linemen. These employees perform hazardous work on electrical lines, often in remote locations and always in teams of three. The behavioral safety process created for these employees involved a combination of self- and peer-sampled observations.

Instead of developing a traditional safety checklist, the linemen created a *lineman index* consisting of a list of safe acts necessary to perform each job safely. Each act was assigned a point value. During an observation, each team received a certain number of points for each act based on how safely each was being performed. For example, one act was described as "outrigger down and pads used" and was assigned 10 points. If the team had the outrigger down but did not use pads, they received 5 of the 10 points. The points received were totaled and a percentage was calculated using the following formula: number of points received divided by total number of assigned points multiplied by 100 percent. These percentages were calculated and graphed on location immediately after each observation.

Now that a clever intervention had been developed, how could we get it off the ground? We faced two significant challenges. First, how would we establish random sampling of observations? Second, how would we motivate employees to perform the observations? Our solution to both challenges was both simple and extremely effective. A fishbowl full of slips of paper was placed near the exit of the linemen's station. Written on some slips was a specific time while others were blank. The times shown on the slips indicated when employees should conduct an observation on themselves and their team members. As employees exited the station to head toward a work site, they were asked to take a slip out of the bowl. To motivate participation, they would automatically receive a treat, such as a candy bar, when they did so. The response was extraordinary! Everyone wanted to participate. Employees began to set their wristwatch timers to the time indicated on their slips as a reminder to conduct safety observations, and they offered to pitch in money in order to increase the selection of treats.

Every Friday a group meeting was held to celebrate the safety process. Celebrations were twofold: (1) each team graph was recognized, simply for the time having been taken to conduct observations, and (2) any increases in safety that occurred throughout the week were recognized and praised. The process was actually somewhat dependent on supervisory involvement also. The process was more successful for supervisors who were more positive and encouraging. Teams whose supervisors inquired frequently about their safety progress usually experienced the highest levels of participation and safety.

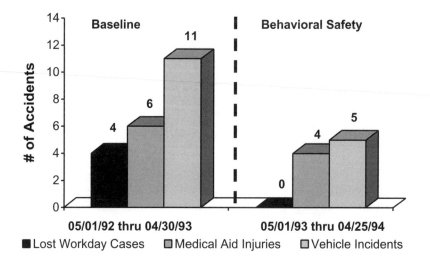

Figure 25.4. Self-observation behavioral safety process in the pilot area at an electric utility helped eliminate lost-workday cases and significantly reduce minor injuries and vehicle incidents.

The power and light company experienced substantial safety improvements as a result of our innovative behavioral safety implementation. Figure 25.4 demonstrates the significant decreases in lost-workday cases, medical aid injuries, and vehicle incidents after the behavioral implementation. Figure 25.5 illustrates the increase in safety scores after the intervention was implemented. The safety committee did not announce that employees were being observed while they collected the baseline

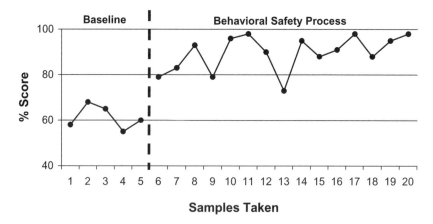

Figure 25.5. Self-observation safety scores before and after implementation.

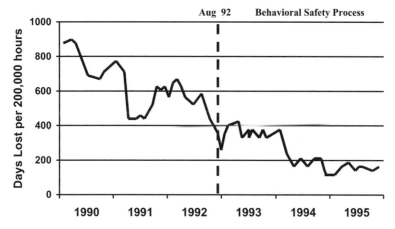

Figure 25.6. Decrease in disabling injury severity after safety intervention at an electric utility.

data. Baseline safety performance that had averaged 60 percent increased to an average of 90 percent after the implementation of the behavioral safety process.

Initially, safety implementation with the linemen was simply to be a pilot study, but as a result of word-of-mouth, maintenance personnel chose to implement the behavioral safety process in their area as well. These employees were responsible for repair of the trucks used by the linemen. When word got around that the linemen were having so much fun and receiving treats and plaudits, they too wanted to participate. Figure 25.6 illustrates the substantial decrease in disabling injury severity that occurred throughout the entire organization as a result of a behavioral safety intervention in all areas. Due to the transportability and ease of use of the intervention developed for the linemen, the maintenance department was able to implement its own safety process quickly and effectively.

Clearly, some organizations require only a quick, easy, and creative safety program in order to achieve dramatic positive results.

25.3 LOGGING INDUSTRY

Some organizations have a very tense and negative environment due to the extremely dangerous nature of the work involved and therefore do not welcome peer observations. Such was the case with an organization in the logging industry.

Loggers and fallers are responsible for cutting massive-sized trees and harvesting immense amounts of wood. They work in isolated wooded areas surrounded by extremely hazardous conditions and *always* work with a partner. An initial examination would suggest a peer-to-peer observation system would best fit an organization without lone workers, but further investigation revealed otherwise. First, relationships among employees were very tense and negative. Second, the organization was located in a state where employees would not be covered by disability if it were found they were responsible for the incident. Therefore, employees were extremely fearful and reluctant to participate in any program that would document their safety performance and could possibly link them to a hazardous action or, worse yet, a severe incident. As a result, employees took *extreme* caution in developing a data collection process that would ensure their anonymity and did not include peer-to-peer observation.

We began our consultation services with two divisions within the organization. Division 1 was comprised of fallers, those who worked up in the trees, and division 2 was largely made up of loggers, those who worked with the lumber on the ground. The implementation process was very similar for each division and began with a two-day training session attended by all management and a representative from each crew. As part of the training, representatives developed a sample safety index for their crews. Each index listed safe behaviors that were essential to performing their job safely. Next, all employees attended a 3-hour "overview" session that outlined the safety process. At the end of the session, each crew elected whether or not to participate in the safety process. If the crew chose to participate, the safety consultant met with them to help build a specific safety index using the sample index as a guide.

The safety representatives (employee volunteers) received more specific training concerning the rollout and maintenance of the process. First, the safety "reps" piloted the safety checklists and conducted several observations, thus allowing us to obtain baseline data. Next they were trained on how to graph the safety data and reinforce participation and improvements in safety. Due to the existing tense environment, any amount of participation was initially reinforced. Unlike division 1, division 2 experienced a great deal of resistance to the safety process. Only 2 of the 30 employees initially volunteered, but by the end of the first quarter, 28 employees had volunteered to assist with the safety process. The specific reinforcement techniques used by the division 2 safety representative are discussed in further detail below.

Although the development of the safety indexes and management of the process were the same across each division, the observation system varied for each. Fallers initially engaged in peer prompts and self-observation. In other words, each employee would prompt their partner to conduct a self-observation using their *faller's safety index*. The actual index form used by the fallers requires some further mention to help explain the caution taken to protect their anonymity. Fallers feared that they would be identified by their penmanship if their indexes required them to "mark off" items. Therefore, they developed a form that required them to punch holes next to each item. Using the end of their tape measurers, employees would punch a hole in the appropriate box for each item. After becoming familiar, and perhaps more comfortable, with the safety process, employees realized that their indexes were missing some critical ergonomic practices. Therefore, they developed another index that was then printed on the back of their existing index. At this time,

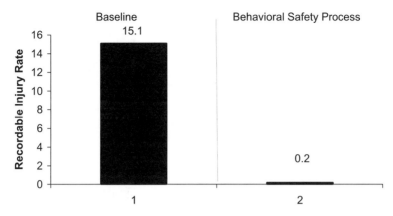

Figure 25.7. Recordable injury rates before and after implementing a behavioral self-observation process in division 1 of a logging operation.

division 1 moved into a combination of self-observation and peer-to-peer observation system. Employees realized that it was very difficult to self-observe the ergonomic items on the checklist and asked their peers to conduct observations on these items. The results of their unique observation system were substantial. Figure 25.7 displays division 1's recordable injury rate (RIR) before and after the behavioral safety intervention. Two quarters after the start of intervention the RIR decreased almost 15 points.

Division 2 employees were prompted to conduct self-observations by three loud blasts emitted from the loudspeaker system. At this time, all employees would take a moment to conduct an observation using their specific index. The indexes were dropped off during their coffee break. The safety representative responsible for this division was faced with an extremely low participation rate and resistance to the process and therefore came up with creative ways to reinforce participation and safety improvements. One day, the safety rep went fishing and traded in his catch for some homemade pies. He brought in the pies to celebrate and encourage participation. Another time, the safety rep brought in a cake during a coffee break. The cake was decorated with a graph depicting their safety performance. These creative, personal, and clearly effortful reinforcers played an important role in creating a more relaxed and positive culture, particularly regarding safety.

Figure 25.8 shows division 2's average safety and participation percentages after implementation. Note the initial drop in safety percentages after the start of publicly posting the scores. This drop is attributed to an initial inflation of scores. Employees reported being concerned about the public posting of data and inflated their scores. They felt more comfortable with the design after they saw the graphs and were able to observe everyone's reaction to the data. Thus, employees became more truthful when conducting self-observations and their safety scores slightly dropped. Another note should be made concerning the two data points that have an arrow pointed

232 SELF-OBSERVATION CASE STUDIES

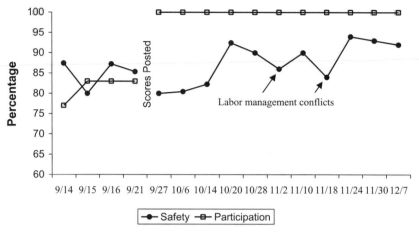

Figure 25.8. Division 2 — safety and participation percentages.

toward them. These two points represent specific dates when the organization experienced union and management problems. Employees reported being upset and distracted on these specific dates and thus did not perform very safely. Figure 25.9 represents the substantial decreases in RIR after implementation of the behavioral safety process.

Sometimes the best way to truly understand the impact a safety process has had on an organization's culture is through the actions and comments of its employees: One employee who had been in the logging industry for 35 years was vehemently against the behavioral safety process. Until he conducted his first self-observation, he claimed there was no need for an observation system. Then he stated he was surprised at how much he had been doing unsafely. He became a safety

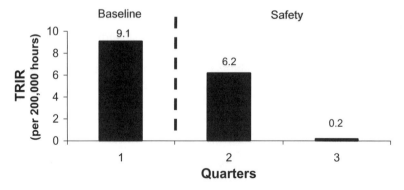

Figure 25.9. Division 2 achieved significant reductions in its total recordable incident rate (TRIR) after implementing a behavioral self-observation process.

representative and now encourages others to do the same. A few years after implementation, a fatality occurred with a faller in another area of the company that had not implemented a behavioral process. The employee, who had once been so opposed to the behavioral safety process, was most upset about the incident and volunteered to implement the safety process in the two other divisions. He was successful with his implementation efforts and continued to be a strong supporter for many years.

26 Small-Company Case Studies

A question that often gets asked is whether behavioral safety is appropriate for small organizations. Often the challenge is one of figuring out an implementation strategy that will work in such organizations. The following case studies show the potential effectiveness of a behavioral approach when the logistics can be addressed effectively.

26.1 PIPELINE COMPANY[1]

Monitoring the safety of employees who work in small isolated groups poses a particular challenge for safety improvement efforts. Such was the case for an oil and gas company with which we consulted. Its pipeline maintenance employees worked in field groups of only a few people; their job was to maintain miles of pipeline in the so-called middle of nowhere. These workers were at risk not only for physical torque injuries and line-of-fire injuries from the energized systems they worked with but also for injuries from natural hazards (such as snakes) in the rough and remote territory in which they worked.

A team of seven workers from three different field crew locations and the district manager designed the safety process. They created a checklist that worked for all of the field groups and an additional one for clerical and office settings. The field checklist included items relating to body position, line of fire, general work conditions, and vehicle safety.

The safety process was designed much like the process detailed throughout this book and included peer-to-peer observations. All employees were asked to conduct one observation per month. Managers and supervisors conducted observations twice per month. Data were sent from the field locations to the steering committee for review and problem solving. Summaries and trends were sent back to each location for field crews to discuss and plan their own safety action plans. In addition, locally planned celebrations were conducted based on the number of observations made, percentage of employee participation, and team achievements involving safety (there were no formal incentives based on accident statistics). To aid communication, both a newsletter and an online system were utilized that provided information about safety accomplishments and listed ideas and

---

[1] This section was written by Judith E. Stowe and Terry McSween, of Quality Safety Edge, and Angelica Grindle, of Western Michigan University.

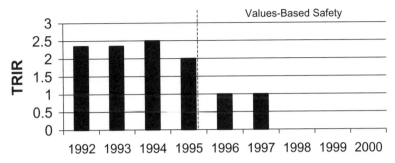

Figure 26.1. After implementing a behavioral safety process, this pipeline company achieved significant reductions in its total recordable injury rates (TRIRs) and experienced zero injuries for three consecutive years.

suggestions for improving safety. After planning these elements of the process, the Quality Safety Edge (QSE) consultant conducted 4 hours of observer training for all employees across a three-day period at three pipeline locations across the state. Immediately following training the teams held kickoff barbecues at each location to celebrate the startup of observations. After the safety process was implemented, in 1996, three years passed without a lost-time injury. It was still in place and functioning effectively as of 2000, when the company was reorganized as a result of a merger. The improvement in safety is shown in Figure 26.1.

A critical factor in the success of the QSE process was the strong support and involvement of the district manager and the development of trust between hourly workers and management. Given the nature of their jobs, people often relied on each other and therefore were completely open with each other about safety matters. As a result, all employees embraced the QSE process and were truly proud of their safety performance, as evidenced by the fact that, although this was a voluntary system, everyone conducted observations. Because of their success, the QSE methodology was also adopted at another pipeline district working with one of our competitors.

26.2 POLYOLEFIN PLANT[2]

Obviously, concern about safety exists in small as well as large plants. However, small plants often do not have the resources to devote to an extended safety assessment and design phase. This was the case in a 100-employee plastic extrusion plant with which we consulted. To meet this plant's needs, we worked with a six-member design team in the planning of its safety process and kicked it off within six weeks. The safety process was designed as discussed throughout this book,

---

[2] This section was written by Judith E. Stowe and Terry McSween, of Quality Safety Edge, and Angelica Grindle, of Western Michigan University.

primarily a system of employee-conducted safety observations together with incentives for individual participation, including such small items as hard-hat stickers, koozie cups, and coffee mugs. Team and department process milestones were recognized with names on a plaque, inexpensive logo watches, and so on.

Shortly after the safety process was begun, some departments reached 97 percent participation from trained observers, and incentive milestones were rapidly attained. However, despite this overwhelming level of participation, the expected safety performance improvements did not materialize. After careful examination we found two main problems. First, upper level managers were pressuring supervisors, who in turn were pressuring hourly workers, to conduct observations, thereby effectively eliminating the concept of "voluntary" observations. Second, teams and departments were competing against each other to be the first to receive the seemingly modest incentives. The solution to these problems was twofold. Once the plant manager was alerted to the consequences of his overly strong safety message, he instructed midlevel managers to stop pressuring employees to conduct safety observations. In addition, the incentive system was modified to eliminate any potential competition between employees. Instead of group rewards, one dollar went to a charity for every observation conducted, with employees nominating the charity after conducting an observation. The charity with the most nominations at the end of the month received the incentive money. The observation process got an added boost when charity representatives started coming into the plant to thank the employees. With these modifications participation in the process dropped to about 60 percent but, interestingly, safety performance improved. Steering team personnel reported that while employees were turning in fewer observation checklists each month, those checklists were higher quality. In other words, data from the observations were more likely to reflect activities observed in the workplace and less likely to be incentive-driven fabrications. The safety process resulted in improved safety performance, as shown in Figure 26.2. The average annual number of total recordable unsafe incidents decreased from 17.9 prior to the implementation of the safety process to 3.4 after implementation.

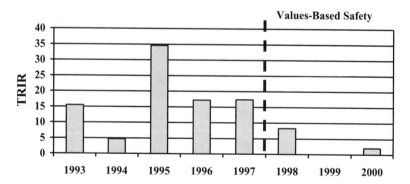

Figure 26.2. This polyolefin plant achieved significant reductions in total recordable injury rate after implementing a behavioral safety process.

As with most successful behavioral safety processes, this one continued to evolve with the changing needs of the facility. The safety committee and management decided that participation in observations would become a job requirement and ended the monthly incentive program. Nonetheless, performance improvements continued. Does this suggest that incentives were not needed in the first place? No. The incentives served crucial purposes in that they (1) helped promote and sustain a high level of employee involvement, (2) increased motivation concerning safety, and (3) helped overcome initial resistance to the process. Over time employees lost interest in the charity donation program, which underscores the importance of keeping incentives fresh and tailoring them to the individual, as discussed in Chapter 11. Fortunately, by the time this occurred, conducting safety observations had become a job expectation at this plant, and participation continued to support a consistent level of safe work practices.

26.3 FOOD-PROCESSING PLANT[3]

This case study shows how the successful implementation of a behavior-based safety process can inadvertently and positively affect other aspects of an organization, in this instance ranging from increased attendance to changing a negative attitude about safety.

In 1999 major rice manufacturer sought assistance in implementing a value-based safety process (VBSP) at their Louisiana plant. Three hundred employees worked in two main areas: the mill, packaging, and shipping area and the trucking, receiving and drying area (the rice drying takes place in heated silos). The trucking area manager did not wish to participate in implementing a behavioral safety process. Therefore, implementation was initiated in the mill area, and by default, the trucking area served as a control group. The VBSP implementation followed the model presented in this book. Before it was implemented, the mill area reported up to seven lost-time incidents per month. Within the first six months after implementation, a 50 percent reduction was achieved in lost-time incidents, while the control group reported three major lost-time incidents (Figure 26.3). The dramatic success of the VBSP in the mill area, along with continued occurrences of major incidents in the control group, helped the at-first reluctant trucking area manager become a strong supporter of the behavioral safety process. Thus, the VBSP was soon implemented in the trucking area as well.

As a result of successful implementation of a behavioral safety process, the Louisiana plant experienced a dramatic improvement in employee attendance. Previously, employees felt the organizational culture was very negative and management driven, thus contributing to the low attendance rates. The employee-driven safety process helped create a more positive work environment and the site experienced noticeable improvement in attendance. Management also recognized

---

[3] This section was written by Wanda Myers and Terry McSween, of Quality Safety Edge, and Alicia Alevero, of Western Michigan University.

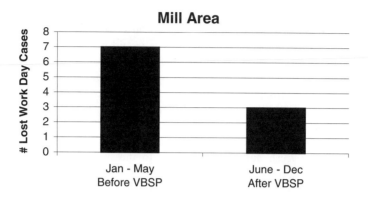

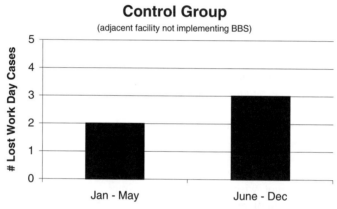

Figure 26.3. This rice-processing plant reduced lost-workday cases by 50 percent in the first six months while an adjacent area that elected not to participate in the behavioral process showed no improvement.

the value and effectiveness of the VBSP concepts and requested assistance in implementing these concepts to help solve other work performance concerns.

In 2001 we were asked to assist a Texas plant with its struggling behavior-based safety program because of the significant safety improvements achieved in the Louisiana plant after VBSP implementation. This plant has approximately 250 employees and deals mostly with instant rice. Another consulting firm had assisted with the initial behavior-based safety implementation at this location. But after four years, the site was experiencing three problems with their process: (1) participation—defined as conducting at least two observations per month—was down to only 10 percent, (2) the steering committee was ineffective, and (3) the process had not reduced injuries. The steering committee was not effectively addressing safety issues and was not effective in promoting participation in the process. These problems were closely related in that employees did not participate because nothing was being done with the data. It was not used effectively for feedback, or to identify and address safety issues, or as the basis for recognition and celebrations. In short, employees had no reason to participate in the process. In addition, management had no formal role in the original process. Once these issues were addressed, participation rose significantly, and both management and employees believed the process was functioning much more effectively. Furthermore, this site achieved a 65 percent reduction in injuries during the year following these enhancements.

27 Observer Effect

In the first edition of *Values-Based Safety Process*, author Terry McSween wrote, "no one has yet conducted definitive research into the effects of employees conducting observations in their own work areas" (p. 24). We found this statement interesting and worthy of further investigation. On discussing the matter with other researchers and practitioners in the field, we decided that enough general interest existed to pursue additional research of the observation process. Many practitioners we approached felt that employees who conduct safety observations on the behavior of co-workers work more safely as a result of conducting those observations. However, this claim had not yet been scientifically tested. Both the feedback and observation processes are critical components of behavior-based safety (BBS), but interestingly, there are many studies on the effects of feedback, whereas studies of the observation process are almost nonexistent. These factors made our task both more interesting and even more challenging. So with these considerations, we began our search to determine if employees who conduct safety observations do work more safely as a result of *conducting* these observations. We termed this hypothetical effect the *observer effect*.

Our first step was to determine whether to conduct the research in a real-world setting (an actual business) or in a laboratory setting. Ideally, we would have preferred a real-word setting, but practical advantages pointed toward the laboratory. First, we felt it would have been difficult to convince an organization to spend extended time and resources for the sole purpose of testing for the existence of an observer effect. Employees would need to be taken from their work to be trained as observers, and unless the organization was interested in implementing a behavioral safety process, it would have been difficult to get the employee buy-in that is so critical to the success of BBS. Further and most importantly, in a laboratory setting we would be able to eliminate the extraneous factors that would be present in an organization and therefore would be better able to attribute the effects, if any, to the intervention and not some other event within the organization. In this case, before trying this in an applied setting, we wanted to be sure that the behavior change of the observer was entirely due to conducting observations of co-worker behavior. Based on these factors, we decided to begin our observer effect research in the laboratory.

Our university laboratory facilities are made up of two empty rooms both provided with a video camera mounted in the top left-hand corner of the room and a

Note: This chapter was written by Alicia M. Alvero and John Austin, of Western Michigan University.

control room furnished with two televisions and VCRs one of each connected to the empty laboratory rooms. For the purposes of our research, we furnished these two rooms with the following materials in order to simulate an office setting: a desk, chair, computer, telephone, cardboard box, and bookshelf. Both of the research studies discussed in this chapter were conducted in these facilities.

For our first study we recruited 12 undergraduate students to serve as participants. Each was randomly assigned to one of two groups, A or B, and given a list of instructions that described the work they would be asked to perform during the experiment. The work tasks were chosen to simulate the work of an office setting and involved lifting light objects, using a telephone, and typing. Participants were also informed that each session would be videotaped so that data could be collected at a later time. Throughout all phases of the study we measured safety performance on the following eight behaviors: (1) back posture and (2) knee position during lifts; (3) neck and (4) wrist position while typing; (5) back, (6) shoulder, and (7) feet position when sitting; and (8) neck position when using the telephone.[1]

During the baseline condition (when no intervention was in effect), participants simply performed the office tasks described on the instruction sheet, and data were collected on their safety performance for the eight cited behaviors. The next phase of the experiment was an information phase. Participants were given a list of four behaviors that included a description of how to perform each one safely (group A participants were given the descriptions for behaviors 1 to 4 and group B for behaviors 5 to 8), and they were told these behaviors would be measured throughout the study. This information on the four safety behaviors was given to them before every session during this phase. We decided to present them with the descriptions during a phase separate from the baseline phase to ensure that any effects we observed during the observation phase could be attributed to conducting observations and not to participants "figuring out" the safe way to do things. In other words, we wanted to ensure that any changes in safety performance were not because participants learned, from the descriptions provided on the safety checklist, how to be safe. Interestingly, participants did not change their behavior just from knowing the definitions of *safe* and *at risk*: Safety performance did not increase significantly for most behaviors during the information phase.

Overall safety performance averaged 13.1 percent for group A and 10.1 percent for group B during this phase, increases of 6.4 percent and 0 percent, respectively, above the baseline averages. The position of the feet while sitting was the only behavior that increased significantly during the information phase. During baseline, overall feet position averaged 8.3 percent and increased to 84.1 percent for group B, consisting of the six participants who were presented with this safety description during the information phase. We speculate that this increase occurred because placing one's feet flat on the floor is perhaps the most easily understood and least effortful of the behaviors to perform safely.

---

[1] For complete definitions contact John Austin, Western Michigan University, Department of Psychology, Kalamazoo, MI 49008; email: john.austin@wmich.edu.

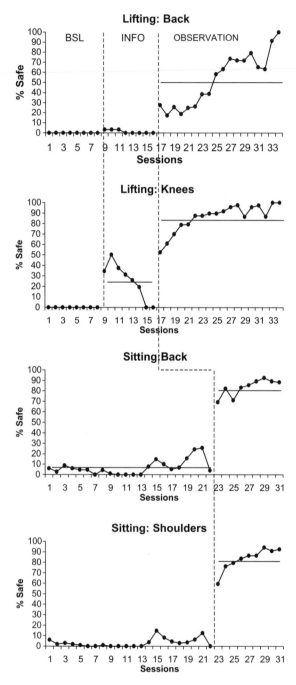

Figure 27.1. Safety performance data averaged across all participants in group A.

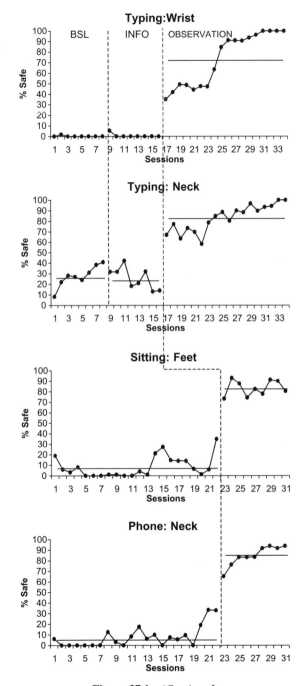

Figure 27.1 (*Continued*)

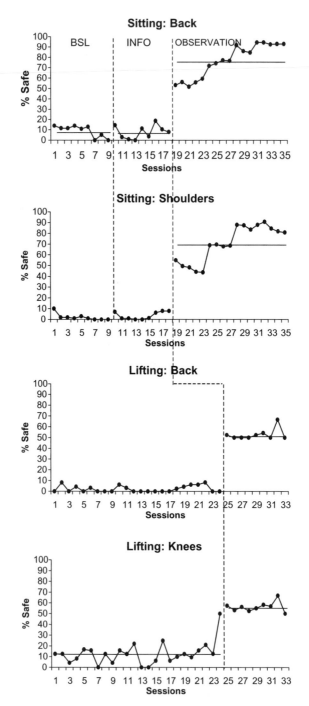

Figure 27.2. Safety performance data averaged across all participants in group B.

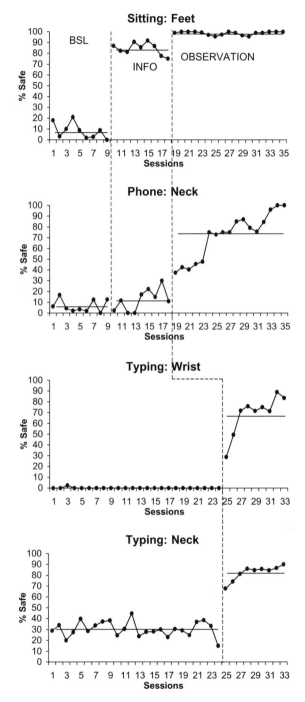

Figure 27.2 (*Continued*)

BEHAVIORS	GROUP A			GROUP B		
	Baseline	Information	Observation	Baseline	Information	Observation
Back during lifts	0	1.2	53.1	2.2	n/a	52.9
Knees during lifts	0	24.9	85.9	12.7	n/a	56.2
Wrist while typing	0.2	0.7	73.6	0.1	n/a	68.5
Neck while typing	27.5	25.5	83	30.4	n/a	82.3
Back while sitting	6.9	n/a	83.4	8.9	7.8	76.9
Shoulders while sitting	3.3	n/a	83.3	2.1	3.6	70.6
Feet while sitting	8.3	n/a	83.3	8.3	84.1	98.6
Neck during phone use	7.8	n/a	84.7	6.7	12.2	71.7
OVERALL	6.7	13.1	77.2	10	10.1	74.4

Figure 27.3. Safety performance averages for groups A and B.

During the observation phase, participants were asked to conduct safety observations while viewing a video of an experimental confederate (an "actor" who was instructed to perform the exact tasks the participants had been asked to perform in the same simulated office setting solely for the purpose of creating this "scoring" video). Participants were given an observation sheet containing a list of behaviors, descriptions of how to perform each behavior on the checklist safely, and columns to score each behavior as safe or unsafe. Safety observations were conducted immediately before participants began their "work" for the session. The observation sheets first given to each participant listed only the four behaviors for which they had received descriptions during the information phase (group A: behaviors 1 to 4; group B: behaviors 5 to 8.) The remaining four behaviors were added to the observation sheet at a later time. This allowed us to study the impact that conducting observations has on the behaviors that were not included on the observation sheets.

The effects of conducting safety observations on safety performance were substantial. Overall safety performance during the observation phase averaged 77.2 percent for group A and 74.4 percent for group B, increases of 70.5 percent and 64.4 percent, respectively, above the baseline averages. Figures 27.1 to 27.3 represent the safety percentages for each behavior averaged across all participants and phases. Figures 27.1 and 27.2 show the safety performance for groups A and B, respectively. Figure 27.3 lists (a) the overall safety averages for each phase and (b) the safety averages for each group by behavior and phase.

Our second study[2] on the observer effect compared the effects of conducting observations to those of feedback. The powerful effects of feedback have often been documented in both laboratory and applied settings. We wanted to know how the effects of conducting observations would compare to those of receiving feedback in our simulated office environment.

We recruited eight participants for the research. Four were assigned to the observation group and four to the feedback group. The setting and procedures for this study were identical to those of the study described above, with the exception

---

[2] This research was funded by the Cambridge Center for Behavioral Studies Safety Grant.

of the feedback group. Participants in the feedback group were exposed to the same baseline and information phases as was the observation group. During the feedback phase, participants received feedback, presented as written percentages, on their safety performance from the previous session immediately before they began each work session. The description of how to perform each behavior safely was included on the feedback forms to ensure that feedback group participants had the necessary information to improve, or maintain, their safety performance. In addition, to ensure a well-designed study, we wanted to treat both groups the same, so we provided the descriptions of how to perform each behavior safely to both the feedback and observation groups.

Following are the findings for two participants, one from the observation group and the other from the feedback group, that are representative of the general findings of this study. Figure 27.4 shows the safety performance for the observation group participant and Figure 27.5 the safety performance for the feedback group participant. Figure 27.6 lists safety performance averages for each behavior across each phase and includes the overall safety average for each phase. Information alone had the strongest effects on behaviors related to lifting for both participants. The effects of conducting safety observations and written feedback on safety performance were sizeable and very comparable. Overall safety performance averaged 86.7 percent during the observation phase and 88.9 percent during the feedback phase, increases of 77 percent and 68.1 percent, respectively, above the baseline averages. The results of this study seem to suggest that the effects of conducting observation are similar to those of written feedback within the context of a laboratory setting.

The results of these research studies provide preliminary evidence for the observer effect. In other words, the data indicate that in a laboratory setting persons who conduct safety observations do perform work more safely as a result of conducting observations.

It is important to note, however, that for several reasons we must exercise extreme caution when interpreting these data. First, these studies are the first to scientifically isolate and assess a main component of the observation process. Therefore, additional data and analysis of this effect are necessary to come to truly reliable conclusions about this component's possible implications for a behavioral safety process. Second, because we isolated the observation process, we are unable to discuss any possible interactions with other components of the behavioral safety process. Further, the observer effect was found to occur in a laboratory setting that lacked many common variables that may affect safety performance, such as social pressures and productivity demands. Therefore, we are unable to confirm that similar observer effect results will occur in a real-world setting. We do suggest that future research endeavors should attempt to replicate a similar study within an organization, and we have been working on such replication studies.

Caution should also be exercised when interpreting the effects of conducting observations versus receiving feedback. Although each component produced substantial effects on safety performance, we are not advocating the use of one component in lieu of the other or implying that one is more critical than the other.

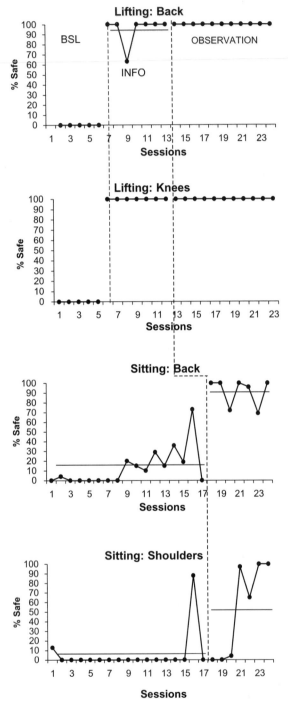

Figure 27.4. Safety performance for observation group participant.

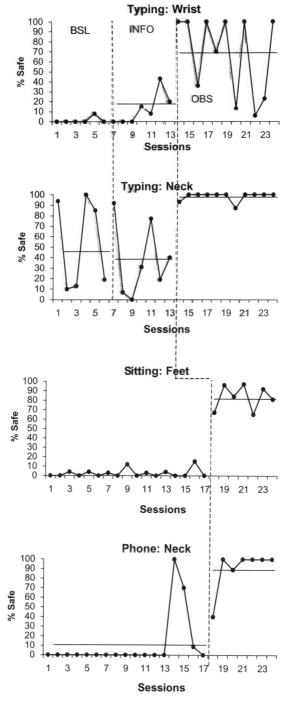

Figure 27.4 (*Continued*)

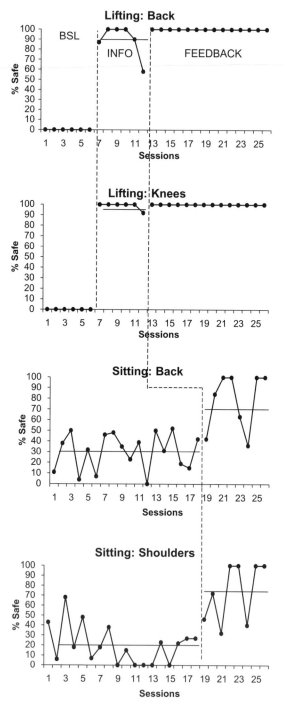

Figure 27.5. Safety performance for feedback group participant.

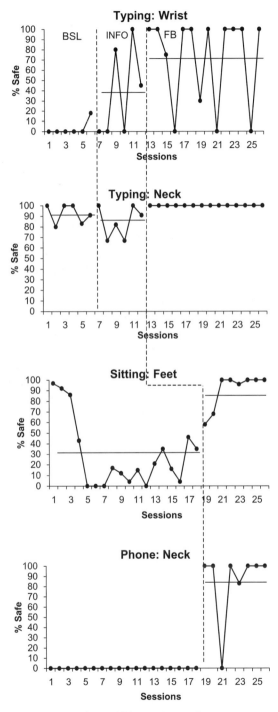

Figure 27.5 (*Continued*)

BEHAVIORS	OBSERVATION PARTICIPANT			FEEDBACK PARTICIPANT		
	Baseline	Information	Observation	Baseline	Information	Feedback
Back during lifts	0	94.7	100	0	89.2	100
Knees during lifts	0	100	100	0	98.7	100
Wrist while typing	1.6	12.3	68	3	37.5	71.8
Neck while typing	45.4	38	98.2	92.3	84.5	100
Back while sitting	13	n/a	91	30.1	n/a	78.1
Shoulders while sitting	5.9	n/a	52.3	20	n/a	73.8
Feet while sitting	2.7	n/a	83	29.1	n/a	90.3
Neck during phone use	10.5	n/a	89.9	0	n/a	85.4
OVERALL	9.7	61.3	86.7	20.8	77.5	88.9

Figure 27.6. Safety performance averages for feedback and observation group participants.

Nevertheless, these findings are interesting indeed in suggesting that it may be beneficial to train all employees to be safety observers.

Further research will be conducted to determine why the behavior of an observer changes after conducting observations. One thing that these data suggest, however, is that the effect is more pronounced for some types of behavior than for others. We would guess that the effects of observing will be more pronounced when the observer is scoring hard-to-judge and hard-to-perform behavior, such as posture, lifting, and other ergonomically related behavior. A behavior such as hard-hat usage would be likely to produce a smaller observer effect, just because the observer already knows the rule of hard-hat usage (wear it or dont wear it). Seeing someone else wear a hard-hat does not give the observer additional information about how to perform the behavior. However, actually seeing a person engage in a particular posture goes a long way toward clarifying the definitions discussed during training, and therefore it produces a larger observer effect.

28 Original Case Studies

The case studies presented in this chapter demonstrate the range of applications of the behavioral approach to safety available when the first edition was being written in 1993. In addition, they also demonstrate some of the early challenges and important considerations. The first two are case studies from this author's files, while the last three are formal research studies conducted by academic researchers working with industrial clients. The first is a good example of the value-based implementation process described throughout this book. It illustrates the usefulness of the value-based approach in building support for a behavioral process implemented in geographically disperse locations and when the logistics prevent having all sites represented in the initial design.

The next two case studies show the potential success of the process yet illustrate the problem of maintaining the process for more than one or two years. In the second case study, the behavioral process dropped out but was subsequently reinstituted by a team responsible for overseeing the process and illustrates the "false starts" that often characterize long-term success. (Remember suggestion 7: Persevere. Don't quit—ever!) The third case study details implementation in a paper mill that lasted two years, after which the director of safety retired and the observations stopped occurring on a regular basis. This case study illustrates common problem when an individual manager or employee champions the process without it being well integrated into the management system. The simple lessons described for maintaining such systems deserve special attention by anyone implementing a behavioral safety process.

The fourth case study illustrates the application of behavioral concepts in addressing industrial hygiene, in this case exposure to styrene. This research study shows the use of observation and feedback to strengthen skills provided through video tape training and suggests the importance of continuing to provide such feedback to ensure a high level of safety practices. The research study is unique in documenting both behavior change and the resulting reductions in styrene exposure.

The final case study demonstrates the effectiveness of safety awards that are small enough to provide a source of motivation but not significant enough to generate false reporting. With all of today's controversy about the effectiveness of incentives, I chose to include this study to show very simply that safety incentives work and should remain an important element of a complete behavioral safety process. The problem is not the effectiveness of incentives, but rather how we use them. This well-designed research study shows the long-term effectiveness of small awards provided monthly on both a group and individual basis for going without an injury. It shows that such programs can be effective and long lasting but often need to be more complex than most traditional safety award programs.

28.1 EMPLOYEE SAFETY PROCESS AT A GAS PIPELINE COMPANY

Implementing a behavioral safety process in an organization that has many geographically disperse locations is a particular challenge. At the time of the assessment, this gas pipeline company had five divisions, each with a division office and multiple compressor stations spread across the pipeline from south Texas to New York. Each station had between 10 and 60 employees. In addition, the company had a very good safety record. It was routinely either the best or in the top two or three of comparable companies within the industry.

Implementation of the behavioral safety process followed the value-based model presented in this book. In consultation with the safety department and management, two pilot locations were selected from each of the five division. The design team included a staff safety representative, an assistant division manager, a front-line foreman, and one hourly employee from five of the pilot locations. The design team completed a three-day workshop on how to design a behavioral safety process, then developed detailed plans for implementation during three two-day meetings. The observation checklist was similar to the one presented in Figure 9.2. The design team named the process the employee safety process to emphasize the employee's responsibility for the success of the process.

After the planning was completed, site safety teams from each pilot location completed a three-day workshop on company values and how to implement the employee safety process. The site safety teams included the site superintendent, foremen, and one or two employee representatives depending on the size of

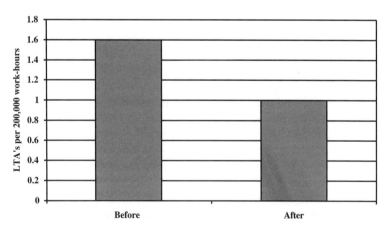

Figure 28.1. Data showing reductions in lost-time injuries at pilot locations of a gas pipeline company.

the location. They were then responsible for conducting kickoff meetings with employees, initiating the observations, training other employees in the observation process, and providing recognition and celebrations of success.

Over the nine months of the pilot the employee safety process achieved a 35 percent reduction in lost-workday cases, as compared with the eight months prior to initiating observations. The data are presented in Figure 28.1.

As a result of this pilot, the management decided to roll the employee safety process out to all locations. Members of the site safety teams from each location participated in two-day workshops that were conducted in each division. A total of 11 workshops were required to train all site safety team members.

28.2 CREATING A POSITIVE SAFETY PROCESS[1]

The union employees in a solids-handling area of a large chemical processing plant wanted to improve their safety. The employees had participated in a behavior management seminar offered as part of the parent company's quality improvement effort and saw how the concepts could apply to safety. As a result of the discussions during one of these seminars, they approached management and asked if they could work with Behavioral Consultant Services to implement a behavioral safety process. They wanted to create a positive, employee-driven safety process, adding that they were tired of having safety programs crammed down their throats and of being beat about the head and shoulders with data from management's safety audits.

Employees volunteered to participate on a positive safety team. The team worked with a resource from Behavioral Consultant Services to design and implement a positive safety process. The process included the following components:

- Team members conducted regular safety walk-throughs using structured checklists to evaluate safety practices and conditions in each area.
- Employees and groups received reinforcement for improvement in safe behavior in their work areas.
- Employees used a suggestion system to identify safety concerns related to either procedures or conditions.
- A computerized message center was used to communicate safety issues to employees in each area.
- Safety observation data were reviewed regularly in area safety meetings and each week in the positive safety team meetings.

The positive safety team continues to meet and refine the observation process, which has already identified and resulted in significant safety improvements in the areas. The improvement data are presented in Figure 1.2. The employees have

[1] This section was written by Terry McSween and Wanda Myers, of Quality Safety Edge.

actively participated in the suggestion system and continue to make regular suggestions and identify issues that need attention. The positive safety team continued the observation process for 15 months, then tried to shift responsibility to teams established on each shift. Unfortunately, the observations did not continue. The area had its first recordable incident 18 months after the observations were initiated and 3 months after they stopped being made regularly. After the incident, the positive safety team began meeting again to reinstall the process and to begin the process of bringing new employees into the team.

28.3 LESSONS LEARNED IN ENHANCING SAFETY PERFORMANCE IN A PAPER MILL[2]

This case study describes a behavioral safety program we coordinated as outside consultants at a paper mill in a small New England town over a period of about a year and a half. To understand the challenges involved, it helps to have some understanding of what is involved in the production of such specialty paper products as toweling and decorative napkins.

The manufacturing process consists of a complex set of procedures. Wood pulp along with rag and paper scraps are mixed with other additives in huge vats. The resultant soupy material is squeezed, pressed, dried, rolled into huge cylinders, cut, decorated, and packaged. Within this dank and odoriferous surrounding, people are continuously active as they guide the process: overseeing the delivery of the raw material and its deposit into vats; the mixing, pressing, cutting, transporting, shaping, printing, and storing; and a myriad of other tasks. All provide numerous opportunities for injuries. Floors become wet and slippery in the area where the pulp is prepared and pressed. Unwisely, some workers attempt tasks on their own that call for mechanical or human assistance, as in trying to lift large paper rolls by themselves rather than using a mechanical hoist. Instead of taking the long way around to another work area, employees might try shortcuts by stepping over moving conveyor belts. Sometimes workers aim to simplify a cutting or pressing operation by dismantling safety devices, such as disconnecting one of two switches required to activate the machine. Also, perhaps to relieve the boredom or to attract attention, workers occasionally clown around. At the site of the present study, we saw a young male driver dart suddenly around a corner on a fork-lift, pretending to aim for an attractive young woman. Fortunately he slowed and veered off in time. But what if he had not?

With a workforce of only a few hundred, injury rates at the plant had remained reasonably low, but injuries did occur from time to time. The unsafe behavior that

---

[2] This section was written by Beth Sulzer-Azaroff, of the Browns Group, based on a study by B. Sulzer-Azaroff, C. J. Fox, S. M. Moss, and J. M. Davis, Feedback and Safety: Involving Workers, unpublished manuscript, 1987. The research was supported by a grant (No. 1 ROI-OHO1928-01) from the Department of Health and Human Services, Public Health Service, National Institute of Occupational Safety and Health.

we observed suggested that more incidents were waiting to happen. When someone was hurt, the whole routine was disrupted. Of course, the victim could no longer continue at the job. Others in the vicinity stopped to lend their assistance or just watch. Another employee would need to accompany the injured person to the plant nurse or a hospital, and it would take time for production to resume its normal pace.

If the injury were serious, the worker usually lost many days from the job. Then a replacement had to be hired and trained. Meanwhile, additional work inundated the supervisor, safety manager, and personnel department as they investigated the incident, then prepared and filed the necessary reports. In some instances a hefty compensation award had to be provided. Obviously, if injuries of this sort happened too frequently, insurance costs would increase.

From a humanitarian perspective, worker and management representatives on the safety committee were eager to see steps taken to minimize injuries to employees. Also recognized was the cost of every incident, resulting in total enormous expenditures that would otherwise appear on the profit side of the balance sheet. They therefore welcomed our research team to help design and conduct a plantwide behavioral safety process.

First we tried to learn as much as possible about current safety procedures so that our behavioral safety process would readily complement it. A director of safety and a committee of worker and managerial representatives were to coordinate the system. After explaining our general approach to enhancing safety performance, we enlisted their participation. The committee agreed to help identify sets of safe performances for each operation in the plant. Many employees were interviewed, sections visited, reports read, and publications consulted in an effort to pinpoint desirable performances crucial to worker well-being. These items were listed for each job, either as needed actions or as the results of behaviors that a passing observer could readily see. Included were such items as the use of protective equipment, specifications for the height of storage piles, hand positions while operating machinery, housekeeping guidelines (such as keeping aisles to exits and fire extinguishers clear and floors free of standing water), and many others.

After definitions of the work practices were refined sufficiently to enable any observer to agree whether it was safe or unsafe, the desired safety practices were used to prepare an observation checklist. The checklist for workers responsible for removing paper rolls from the spindle and preparing them for transport to the printing division included such items as "uses hoist when removing roll from spindle." Checklists in the packaging area included "pallets contain piles no higher than six feet" and "area between yellow lines [demarcating walkways] free of any material." Workers who used heat to seal packages were to wear protective gloves, and those in noisy locations, ear protection.

The checklists were tested and revised until scoring was unambiguous. Next, teams were formed of participants in operational units. Whenever a team was observed, it received a "percentage safe" score based on the number of items the observer saw people performing correctly and incorrectly and on the number of areas free of hazardous conditions. The results were transferred onto graphs. This procedure continued twice a week over several weeks until weak and strong patterns of performance emerged.

At this point, the members of some of the teams were invited to participate in the safety improvement process. In a kickoff meeting their pinpointed performances and the rationale for selecting these, the systems of observing, and the results already obtained were explained. Specific data were shown to team members and summary graphs prominently posted in their area. Each team then selected a performance that it realized needed to improve to achieve safety on the job and on which it would concentrate for the next few weeks. Later it would select new performances for emphasis and other teams would join in. Thereafter, whenever a team was observed, a copy of the completed observation checklist and the team score was posted in the work area. Members were commended for any improvement. The observation data and progress toward each target are shown in Figure 28.2.

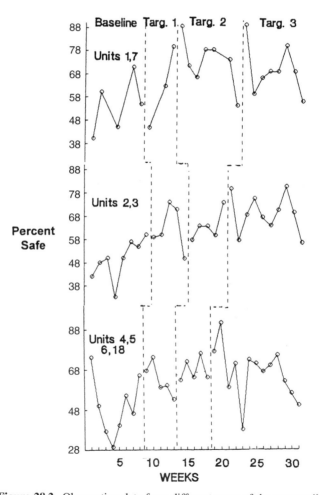

Figure 28.2. Observation data from different areas of the paper mill.

The impact was almost magical. In tandem with the posting of the safety data, performance quickly began to improve. Simultaneously, rates of injuries started to decline. This enhanced performance continued throughout the duration of the program. Here was an illustration of the strength of the behavioral safety process. It worked not only to reduce unsafe acts and conditions but incidental injuries as well. In the process, power was delegated to personnel at all levels. They readily participated in selecting and defining pinpointed performances, assisted with the observation system, chose performances to emphasize, and took part in the observations.

The system remained in effect for two years, by which time each team had elected a member to conduct the observations and provide feedback. But then, unfortunately, the safety observations gradually began to slack off. We were no longer active consultants, and because the director of safety retired and was not replaced, a leadership vacuum developed. No one else was assigned to manage the safety improvement effort. Our follow-up assessments revealed that performance began to deteriorate. It was feared that eventually rates of incidents would increase.

The lesson learned from the lack of long-term persistence of the system was sobering. It did teach us, however, that henceforth any behavioral safety process we helped design would have to be incorporated as a "way of life" within the standard operation. Responsible individuals within the organization would eventually need to assume any ongoing activities that we had initiated as outside consultants; further, ongoing reinforcement would have to be structured for those individuals. For example, in our current work, teams and their supervisors elsewhere typically receive tangible rewards for improving and sustaining good safety performance records while senior managers are periodically awarded bonuses according to their maintenance of the safety system. Such arrangements and incentives do seem needed. Other than the program at the paper mill just described, every one of about a dozen safety programs for which we have served as consultants has lasted within the organization, to the benefit of both workers and management.

28.4 BEHAVIORAL APPROACH TO INDUSTRIAL HYGIENE[3]

Handling toxic substances is a work requirement for employees in many industries. The potential perils of handling toxic substances are recognized by many in those industries, which include the fiberglass-reinforced plastic products industry (plants in this industry, which typically resemble warehouses, produce products such as molded fiberglass boats as well as the bathtub/shower units found in many homes). An important task is to develop ways to reduce workers' exposure to the harmful chemicals, as by calling on engineers to redesign equipment or manufacturing processes. The question remains, however, whether or not these engineering solutions are as effective as they possibly could be because their effectiveness

---

[3] This section was written by Tracy Thurkow and Bill Hopkins, of Auburn University. This case study is based on Hopkins et al. (1986b).

depends, in part, on the workers who use the technology. Workers must actually *use* the technology. If they do not, the technology can hardly be as effective as it could be in reducing their exposure to toxic chemicals.

Workers in the fiberglass-reinforced plastic products industry are required to handle styrene, a suspect carcinogen, on a daily basis. Workers touch styrene with their bare hands, they breathe evaporated styrene from the surrounding air, and they walk on floors covered with material containing styrene. Even though they cannot see the chemical, they can smell its sickeningly sweet smell constantly and everywhere. At levels that occasionally occur in these plants, styrene can make people who are not adapted to it feel groggy and drunk and also burn their eyes. The effects of lower but longer term exposure to styrene are not known.

A research team composed of two behavioral scientists and one industrial hygienist combined their efforts to develop and implement a safety program to limit workers' exposures to styrene. The program was not designed to change the technology in fiberglass-reinforced plastic products plants; rather, it was designed to change workers' behaviors so that they would be more likely to take advantage of the technology already present.

Because the goal was to change workers' behaviors, the research team first had to pinpoint which work behaviors led to styrene exposures and which work behaviors would reduce such exposures. After visiting 15 plants in the industry and observing employees working with styrene, the team determined that behaviors in three specific manufacturing processes, called sprayup, layup, and rollout, resulted in the greatest styrene exposures. Basically, workers who have to perform these processes are required to do the following:

- Spray gelcoat, a combination of pigmented polyester resin and styrene monomer, onto a mold (such as a bathtub/shower stall mold) with a spray gun.
- Spray reinforcing laminations of resin and styrene, with chopped-up strands of fiberglass mixed in, onto the mold.
- Reinforce parts of the mold with pieces of metal, wood, cardboard, or fiberglass mats soaked in resin, styrene, and catalyst.
- Spray the reinforcements with a mixture of styrene and fiberglass.
- Roll the reinforcing layers of resin, styrene, and fiberglass with metal rollers to remove gas bubbles.

In general, these manufacturing processes take place inside or close to ventilated work booths that resemble very large metal boxes; several of the work booths may be scattered throughout the plants. In these booths, a large ventilator sucks air in through an open wall or window and through the exhaust ports, or large air filters, along one wall.

While observing workers manufacture products, the research team noted many instances of harmful behaviors. For instance, workers often worked "downwind" of another worker spraying material containing styrene with the result that as the air was sucked back to the exhaust ports, they would be more likely to breathe in air that contained styrene. Another common scene was overspray (hardened resin and fiberglass resembling a blanket of hardened, white cotton candy filled with prickly icicles) covering the plant's floors and exhaust ports, rendering the ventilator system less useful. Sometimes workers would be so involved in their work while standing in the middle of this overspray that their boots would become stuck as the resin hardened and when they left the area, the boots stayed behind.

Once the research team had observed and pinpointed the behaviors and conditions that resulted in the most styrene exposure, the important task was to identify those that could reduce the exposure. The team consulted the workers who handled styrene during the manufacturing processes for suggestions about how to restructure their work to reduce their exposure to the chemical. The researchers also reviewed more than 250 professional publications for recommendations.

Based on what they heard and read, the researchers developed lists of 10 observable work practice behaviors and 20 specific and measurable housekeeping conditions that could limit workers' exposure to styrene. These lists appear in Figure 28.3.

Next, the team recruited area plants to participate in a safety program. Three plants located close enough to the researchers that used the specified manufacturing processes wanted to participate. We will call them plants 1, 2, and 3.

The program got under way. The research team began to observe workers and collect data on specific behaviors in plants 1 and 2 before trying to change workers' behaviors. The researchers could thereby determine how effective their program was at changing workers' behaviors in that they would be able to compare data collected after the program was implemented to data collected before the program began. In particular, the researchers wanted to observe (1) the number of times each worker performed the 10 desired work behaviors, (2) the condition of each work area, and (3) the amount of styrene to which the workers were exposed.

The researchers observed each worker (nine in plant 1 and three in plant 2) four times a day to determine how frequently he or she engaged in the desired work practices and recorded each occurrence. They also inspected the housekeeping conditions in the areas where the sprayup, layup, and rollout processes took place. Information about the conditions was collected five times every day: at the beginning and end of work shifts as well as during the morning, lunch, and afternoon coffee breaks.

The research team ascertained the amount of styrene in the workers' environment by attaching small tubes filled with charcoal to the end of a hose hooked up to an air pump. The tubes were attached to the workers' clothing near their breathing zones; thus the charcoal absorbed an amount of styrene proportional to the amount the workers breathed during the day. The charcoal tubes were then sent to a laboratory where the amount of styrene was analyzed.

In order to change behavior, a member of the research team, a trainer named Anne, first specified which behaviors she wanted the workers to perform. The workers were shown one of eight videotapes every few days during training meetings. Each videotape highlighted the importance of a few of the specific work

Work Practices List:
1. Turning on spray booth exhaust fans in specified circumstances
2. Keeping breathing zones at least 18 inches from sources of styrene
3. Keeping styrene off the skin
4. Working on the upwind side of sources of airborne styrene
5. Working only in specific areas to take advantage of airflow and exhaust ventilation
6. Avoiding a position in front of a person operating a spray gun
7. Avoiding spraying towards another worker
8. Spraying towards the exhaust ports of the booths
9. Spraying so that most of the spray hit the mold or part
10. Locating molds and parts to take advantage of the airflow

Housekeeping Conditions List:
1. All spray booth filters should be in place
2. Overspray buildup on booth filters should be kept below 1.25 cm in chop spray booths and 0.625 cm in gelcoat booths
3. Floors in work areas should be covered by a disposable material
4. Floor coverings should not be torn or soaked through with resin
5. Resin-soaked debris should be removed from work areas
6. Overspray buildup on booth floors should be kept below 2.54 cm
7. Overspray buildup on rollout area floors should be kept below 1.25 cm
8. Floors and work table tops should be free of resin and gelcoat spills larger than a diameter of 15 cm
9. Overspray buildup on booth walls should be kept below 1.25 cm
10. Spray booth lights should be operational and visible
11. Waste cans should be available in spray booths
12. Table coverings should not be torn or soaked through with resin
13. Work areas should be free of empty chemical containers
14. When work on them is completed, curing parts should be removed from work areas
15. Wheels on mold carriages should turn freely
16. Resin and gelcoat containers should be covered
17. Spray equipment gauges should be visible
18. Acetone containers should be closed when not in use
19. Spray guns and hoses should not leak
20. Work areas should be free of food and drink

Figure 28.3. The 10 desired work behaviors and 20 desired housekeeping conditions.

practices and housekeeping conditions. The meetings were very informal; everyone gathered at the beginning of a work shift to have a cup of coffee, watch the 5- to 10-minute tape, and talk about what they had seen. When workers expressed concerns that the safety practices might slow down production, they were told that the company wanted them to take the extra time; they would not need to work overtime to pick up production. The companies felt that the recommended practices would smooth the work out so that any time lost from the practices would be made up through smoother work procedures. At the end of each meeting, each worker was given a list of the goals he or she would need to accomplish in order to pass an on-the-job test on the work practices and housekeeping conditions covered in the videotape.

Second, trainer Anne provided feedback to the workers several times a day during "checkout" times. She praised workers when she observed them performing the desired work behavior and when their work area met the housekeeping goals. If a worker did not perform the desired work practice, Anne provided corrective feedback to tell the worker WHAT was being done incorrectly, WHY it was important that it be done correctly, and HOW the recommended work practice would solve the problem. After a few minutes, she returned and praised the worker for engaging in the desired behavior.

Anne soon developed good rapport with the workers. For instance, Henry, a strong and wiry, rather mellow man in his mid-forties, was a hard worker and as a result probably had a higher than average exposure to styrene since his job involved spraying resin (containing styrene) and fiberglass onto molds. At first he was quiet about the program and ignored Anne, but after the first few training meetings he realized the intent of the program was to help protect him and he began to warm up. He was subsequently often observed talking and joking with Anne. Henry was also well liked by his co-workers and often socialized with them outside of work, which helped in further developing rapport between Anne and the workers.

After a few days, workers took a test on that week's work practices and housekeeping conditions thus far. They had to perform the desired work behaviors 100 percent of the time they were observed in order to pass the work practice test, and their work areas had to meet 90 percent of the housekeeping goals in order to pass that test. If workers passed, they were congratulated and rewarded with either $5 or $25 for passing the work practice test, depending on the specific work practices being tested, and $25 for passing the housekeeping test. A typical good-natured response by Henry after he passed a test and learned how much his reward would be was, "That'll buy another six-pack for the weekend. Everyone can come to my place, share a beer, and I'll cook up something."

While the training, feedback, and testing were occurring in plant 1, the research team was still observing the workers in plant 2. About three months after the program began in plant 1, it was introduced in plant 2. About two months after the program began in plant 2, observations were started in plant 3. After a month of observations, the program was started in plant 3.

When the behavior change program was implemented in all three plants, the research team found that the workers' behaviors changed. Before the program, workers in the three plants performed the desired work practices between 49 and 76 percent of the time; after the program began, they performed them almost 100 percent of the time (see Fig. 28.4). Before the program began, housekeeping conditions met the goals almost 50 percent of the time, and this percentage was gradually worsening in all three plants during the observation periods. However, after the program began, housekeeping conditions in plant 1 met the goals 85 to 90 percent of the time and in the other two plants more than 90 percent of the time (see Fig. 28.5).

The behavior change program also had effects on workers' exposure to styrene, which, of course, was the overall goal of the program. In plant 1, exposure to styrene decreased by 38 percent; in plant 2, by 73 percent; and in plant 3, by

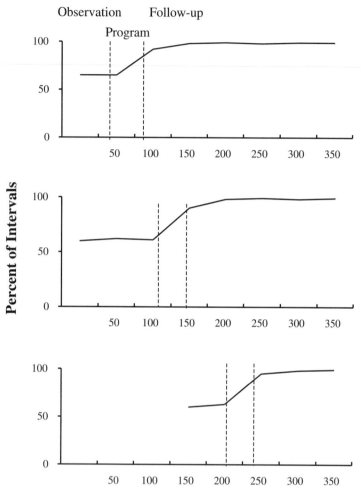

Figure 28.4. Mean percentage of observation intervals in which the workers performed the safe practices.

85 percent (see Fig. 28.6). All of these changes occurred with no decrease in production. The researchers also observed a slight increase in the amount of time workers spent at actual work tasks.

The researchers accomplished their goal of changing workers' behavior, but they were not satisfied with only this. They also wanted to make sure the changes lasted. About two years after the program had ended in all three plants, the researchers were able to check back with one of the plants to see if the safety practices had maintained. They found that several of the safety practices as well as the reduction in styrene exposure had lasted for the two years. Next, the team reimplemented the feedback and safety awards initially used and found the workers responded

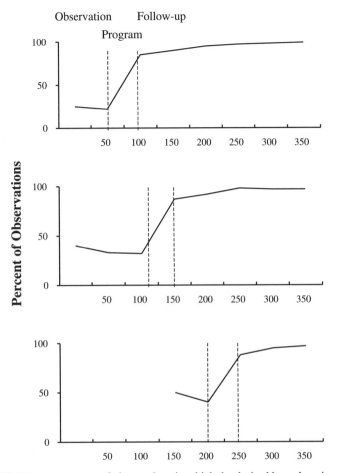

Figure 28.5. Mean percentage of observations in which the desired housekeeping conditions occurred.

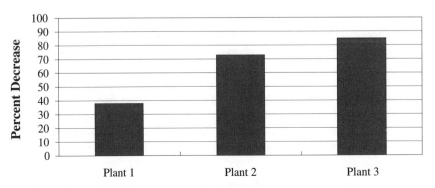

Figure 28.6. Mean decrease in styrene exposure.

immediately. In fact, the frequency of desired safe practices and housekeeping conditions increased when the same behavior safety program was implemented a second time.

This behavior change project demonstrated some important points about behavioral safety programs in general. First, it showed that managing safe behavior can be very important to businesses as well as to employees' well-being. Behavioral safety programs can indeed effectively limit workers' exposures to toxic chemicals, and for some businesses they may be more economical and feasible than technological changes.

The project also demonstrated that this particular type of behavior change program, including lists of specific desired behaviors, training, feedback, and rewards, can be very effective at both changing behavior and maintaining those changes. These changes in behavior can be beneficial to workers and to the businesses in the long run because they can reduce workers' exposures to toxic chemicals by taking advantage of the available engineering technology.

28.5 LONG-TERM EFFECTS OF A SAFETY REWARD PROGRAM IN OPEN-PIT MINING[4]

In the mining industry, safety has always been an enormous concern. At the Shirley Basin mine of Wyoming and the Navajo mine of Arizona, unsafe operations were reaching a critical point in the early 1970s. Both mines had dismal safety records. During 1970 alone, the Shirley Basin mine experienced 104 injuries serious enough to result in workmen's compensation. Moreover, they had 162 vehicle incidents, mostly involving expensive heavy machinery such as dump trucks and front-end loaders. The Najaho mine also had a high injury rate, with two of its incidents between 1970 and 1975 tragically resulting in deaths.

The Shirley Basin mine was a large uranium-ore pit mine that employed approximately 400 workers. The Navajo mine operation consisted of three strip-type coal mines and employed approximately 450 workers. At both mines, temperatures were often extreme, sometimes dipping as low as 40 below zero at Shirley Basin. To make matters worse, gusting winds whisked mining dust about almost continuously. Usually, the only sound present at the mines was the din of diesel engines mixed with the clamor of earth being upheaved. In short, the conditions were not conducive to safety.

The miners were a rugged bunch, about what you would expect in such environments. They worked hard to get the ore out of the ground and were proud of what they did. But they often took shortcuts. They drove the dump trucks down into the mines too fast and sometimes had incidents. They repaired machines without properly wiping the grease off and sometimes wrenches slipped, causing

---

[4] This section was written by Bill Hopkins and Jim Getting, of Auburn University. This case study is based on Fox et al. (1987).

cuts and smashed fingers. In other words, they occasionally wrecked equipment and got themselves hurt. Nobody wanted these things to happen, of course, but they did. Mining was difficult and dangerous, it was thought, and incidents came with the territory.

David Fox, the health and safety engineer at Shirley Basin, was not willing to accept the notion that mining had to be so dangerous. In 1972, he designed and implemented a behavioral safety program at the mine. Fox's assessment of current practices revealed that employees were probably adequately trained and knew which behaviors were safe and unsafe. A health and safety staff trained new employees via classroom and on-the-job training. Mandatory yearly refresher courses kept all employees aware of safety issues. Safety was further emphasized in other ways. Workers who had hazardous jobs took part in biweekly "toolbox" safety meetings where incidents and close calls were discussed and hazardous conditions described to the supervisor so they could be corrected. Additionally, the health and safety staff inspected the entire mine each week in order to identify and minimize hazardous conditions.

Fox decided that more than just training and inspections were needed. In consultation with co-author Hopkins, he designed and implemented a simple incentive system. Each employee was able to earn S&H Green Stamps by avoiding any injury during an entire month. The number of stamps a worker could receive was based on the degree of risk for injury at each job. Four "risk groups" were established with rewards ranging from 300 stamps per month for the lowest risk jobs to 700 stamps for the highest risk jobs. The stamps were redeemable for such common household and personal items as tools, small appliances, and sporting goods. Using these stamps was particularly clever because the health and safety staff did not need to guess which items would be reinforcing to the employees. Instead, workers were able to choose the items they most wanted. Thus, the value of the stamps was maximized. Further, the employees' spouses and children became involved because the stamps could be exchanged to acquire items for family use. These items thereby became constant reminders of safety. For example, an employee who used the stamps to acquire a new gas grill would likely remember that it was a reward for safety each time he or she used the grill.

A group incentive system was also established to complement the individual incentive system. If all workers managed by the same supervisor avoided injury, all were rewarded with more stamps. Again, the number of stamps awarded was based on the risk group to which a worker belonged, ranging in this case from 200 to 700.

Dovetailing a group incentive system with an individual incentive system was an important component of this program because only results were being measured. The actual safe and unsafe behaviors that contributed to the incident and injury rates were not being directly observed or measured. However, the group incentive provided motivation for members of work groups to police each other. Continuous observation was thus in fact also built into the system.

Special bonuses of stamps were also awarded, ranging from 500 to 25,000 stamps. Administered by a safety committee comprised of workers, supervisors, and safety and health personnel, these bonuses were given as rewards for worker suggestions that improved mine safety and to workers who helped prevent a serious incident or injury.

The programs were kicked off by describing them and the award schedules in company newsletters. To emphasize the company's commitment to the new programs, workers were given an initial retroactive bonus. At the Shirley Basin mine, all workers were given 10,000 stamps if they had not suffered a lost-time injury or had an equipment-damaging incident in the past two years. At the Navajo mine, when the program was later implemented there, workers were given 10,000 stamps if they had not suffered an injury or had an incident for the past year.

At first, the miners were skeptical about the programs. When the Green Stamps first began appearing, a few of the Shirley Basin miners were not impressed with the stamps as rewards. However, the Shirley Basin program proved successful. The Navajo mine elected to replicate the program beginning in 1975.

The number of lost-time injuries per million person-hours worked is shown in Figure 28.7. At both mines, lost-time injuries decreased considerably. After program implementation, the rate of lost-time injury at Shirley Basin was about 15 percent of the baseline measure (1970 and 1971) and at Navajo about 32 percent of the baseline measure (1970 to 1974). Comparisons to national averages further portray the effectiveness of the program. Following implementation of the program, the number of days lost to injury at Shirley Basin was about one-fourth the national mining average. At Navajo, it was only one-twelfth the national mining average.

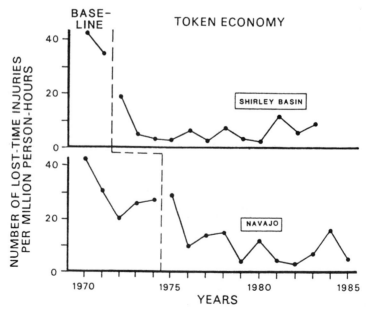

Figure 28.7. Yearly number of work-related injuries per million work-hour requiring one or more days lost from work.

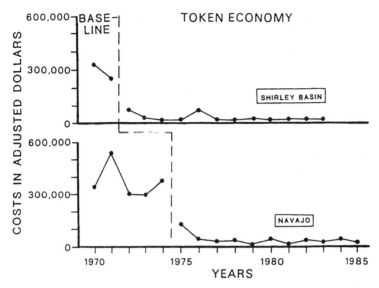

Figure 28.8. Yearly costs, adjusted for hours worked and inflation, resulting from incidents and injuries.

Reductions in injuries were accompanied by reductions in incident rates as well, resulting in considerable savings to both mines. Figure 28.8 shows the yearly costs of incidents and injuries to the mines. At Shirley Basin, cost decreased from a baseline average of $294,000 per year to an average of $29,000 dollars per year. At Navajo, costs decreased from a baseline average of $367,000 per year to an average of $39,000 per year.

The yearly cost of the stamps used in the safety program was easily offset by such savings in reduced incidents and injuries. The cost of stamps ranged from $9000 to $12,500 at Shirley Basin and $11,000 to $13,500 at Navajo. Using a formula that compares dollars spent on the safety reward program to reduction in the costs for incidents and injuries adjusted for hours worked and inflation, these figures translate to yearly returns on investment of between 18 to 1 and 28 to 1 at Shirley Basin and between 13 to 1 and 21 to 1 at Navajo. In short, the stamps paid for themselves many times over. Moreover, fewer people were getting hurt or worse, and the importance of this result far outweighed the monetary savings.

This study shows that behavioral approaches to safety can be maintained indefinitely if the program is well designed. The program at Shirley Basin succeeded for 12 years and would have probably continued had the mine not been closed due to a decrease in the demand for uranium. The program at Navajo was still active and will realize its twentieth year of success in 1995.

The longevity can probably be attributed to the careful considerations that went into the plan. It was known that miners were properly trained and knew how to work safely. Observing rates of safe and unsafe behaviors was judged to be extremely difficult, so the group incentive system was designed to "build in" observation and correction of unsafe behavior by the workers themselves. Also, using the stamps maximized the value of the rewards because workers obtained items they desired and the workers' families became involved with collecting the stamps. Finally, the program offered something for everyone. Workers appreciated getting recognition and rewards for staying safe. Management was saving huge amounts of money. The safety and health staff had many fewer incidents and injuries to investigate as well as an effective safety program that was easy to administer. With care and creativity, behavioral approaches to safety can benefit everybody. The eventual popularity of these kinds of programs is illustrated by the fact that the union at Shirley Basin asked that the stamps program be incorporated into its contract.

APPENDIX A
Sample Implementation Schedules

Figure A.1 is a sample schedule for the safety process with weekly planning meetings. Figure A.2 is a sample schedule for the safety process with two-day planning meetings.

Sample Schedule for Safety Process With Weekly Planning Meetings							
			1st Quarter Jan Feb Mar	2nd Quarter Apr May Jun	3rd Quarter Jul Aug Sep	4th Quarter Oct Nov Dec	1st Quarter Jan Feb Mar
Name	Scheduled Start	Scheduled Finish					
Kick-off Workshop	1/7/93	1/7/93					
Develop observation process	**1/11/93**	**3/19/93**					
Develop list of safe acts	1/11/93	1/22/93					
Revise list	1/25/93	1/29/93					
Draft checklist & process	2/1/93	2/12/93					
Trial run observation process	2/15/93	2/26/93					
Revise forms & process	3/1/93	3/5/93					
Management review	3/10/93	3/10/93					
Kick-off observation process	3/15/93	3/19/93					
Establish feedback and involvement	**3/22/93**	**4/30/93**					
Develop employee safety teams	3/22/93	3/26/93					
Create charts and scoreboards	4/5/93	4/9/93					
Establish data review process	4/12/93	4/16/93					
Management review	4/19/93	4/19/93					
Set improvement goals	4/26/93	4/30/93					
Implement safety recognition system	**5/5/93**	**6/11/93**					
Develop criteria	5/5/93	5/11/93					
Identify potential awards	5/10/93	5/14/93					
Survey personnel	5/17/93	5/28/93					
Finalize plan	5/31/93	6/4/93					
Management review	6/4/93	6/4/93					
Initiate recognition plan	6/7/93	6/11/93					
Provide training support	**6/14/93**	**7/9/93**					
Revise new hire orientation	6/14/93	6/22/93					
Develop observer training	6/28/93	7/9/93					

Project: Safety Process Enhancement
Date: 7/9/94

Critical / Noncritical / Progress / Milestone / Summary / Rolled Up

Figure A.1. Example of a schedule for design teams meeting for 2 hours every week (shows partial implementation).

Sample Schedule for Safety Process With Two-Day Planning Meetings

Name	Scheduled Start	Scheduled Finish
Safety Process Assessment	**2/5/93**	**3/22/93**
Division mgmt interviews	2/5/93	2/5/93
Pilot site visits	2/22/93	3/12/93
Report preparation	3/12/93	3/19/93
Executive presentation	3/20/93	3/22/93
Implementation - Pilot Locations	**4/6/93**	**8/27/93**
Step 1 - Train Design Team	4/6/93	4/8/93
Step 2 - Design Observation Process	5/17/93	5/21/93
Step 3 - Design Kick-off & Safety Awards	6/7/93	6/11/93
Step 4 - Develop recognition plans	6/15/93	6/15/93
Train Site Safety Teams	**6/16/93**	**7/14/93**
Prepare training materials	6/16/93	6/21/93
Initial training - Group 1	6/21/93	6/23/93
Initial training - Group 2	7/12/93	7/14/93
Pilot site kick-off meetings	8/2/93	8/27/93
Implementation Review	11/16/93	11/17/93

Project: Value-Based Safety Process
Date: 7/9/94

Critical / Noncritical / Progress / Milestone / Summary / Rolled Up

Figure A.2. Example of a schedule for design teams meeting for two days every other week (shows complete implementation).

APPENDIX B
Selected Consultants Experienced in Implementing Behavioral Safety Processes

The following is a partial list of consulting organizations that have personnel with at least five years of experience implementing behavioral safety processes. This list is not meant to be all inclusive or an endorsement of the organizations.

Behavioral Science Technology, Inc.
323 East Matillija Street, Suite 215
Ojai, California 93023
(800) 548-5781
www.bstsolutions.com

Marsh, Inc.
3475 Piedmont Road, NE
Atlanta, Georgia 30305-2954
(404) 995-3314

Safety Performance Solutions
1007 North Main Street
Blacksburg, Virginia 24060
(540) 951-7233
www.safetyperformancesolutions.com

Quality Safety Edge
14676 S. Diamondhead
Montgomery, TX 77356
877-588-1140
www.qualitysafetyedge.com

All of these organizations participate in Behavioral Safety Now, a professional conference on behavior-based safety sponsored by the Cambridge Center for Behavioral Studies. Learn more at behavioralsafetynow.com.

APPENDIX C
Unstructured Approach to Identifying and Defining Values

CLARIFYING YOUR VALUES

The phases below outline the original, unstructured process from the first edition that we used to assist clients in identifying values for their process. This approach is included because it may still be an appropriate approach for some organizations and of interest to those who wish to understand the history of the structured process described in the current text. You should consider the following phases if this approach is more appropriate to your situation:

- Phase 1: Identify your basic values.
- Phase 2: Pinpoint practices that exemplify those values.
- Phase 3: Provide training on the new values.
- Phase 4: Use the values as basic ground rules for interactions.

Use values to clarify your team's expectations about how managers and employees will interact with one another within the organization's safety process. For this procedure, the ideal group is a "diagonal slice" of representative employees from different levels, areas, and functions within the organization.

These steps also work well for teams addressing issues other than safety. For example, you might use this process to help establish and clarify your organization's basic values or to clarify the type of personal interactions required to ensure the success of other improvement initiatives.

The following table is a guide for getting started:

If Your Organization:	Then:
Does not have a set of formally defined values	Start with phase 1. Use either a group or individual process for developing a list or values; then define practices in support of those values that will contribute to the success of your safety process

If Your Organization:	Then:
Has formally developed a set of values	Go to phase 2. Define practices in support of those values that will contribute to the success of your safety process

PHASE 1: IDENTIFY YOUR BASIC VALUES

Step 1: Review and discuss your company vision or mission statement and company value statements. Have each member of the group read these statements, either before the team meeting or during the meeting preceding the discussion.

Step 2: Brainstorm the group's values. Brainstorm your organization's values as a group activity during the meeting or ask team members to develop lists of value statements before coming to the meeting.

For a group brainstorming session, list the following questions on a flip chart, then tape the page on a wall or easel so it can be seen by all team members:

- How should you treat others at work?
- How do you want to be treated?
- What words would describe our ideal interactions at work?

To enable team members to consider their values before the meeting, you may want to distribute a memo that explains the task and includes these questions. Then review and record the value statements developed by individual members. List all responses on the flip chart so that everyone can easily see them. Tape each page on a wall as you fill the page.

Step 3: Cluster related practices and combine ideas. Try to group the responses into categories of related items as your list is likely to include several practices that can be captured, combined, or summarized by broader value statements. Create a new condensed list that clearly states each idea. Continue to work on condensing your list until you have five to nine broad value statements. Save statements of specific practices and behaviors for use in phase 2.

PHASE 2: PINPOINT PRACTICES THAT EXEMPLIFY THOSE VALUES

Practices are descriptions of behaviors that operationally define each value statement. The best practices for your organization will describe behavior directly pertinent to the success of your safety process. In addition, they should be worded in language that is meaningful to the people who will see them.

Step 1: Draft an initial list of practices for each value statement. As with the value statements, you may identify practices as a group or individually. This process usually works best when each employee has an opportunity to describe practices that support the organization's values before the team meets as a group.

Ask team members to identify practices or activities that would support each value statement and then practices or activities that would be contrary to each value statement. You should focus primarily on identifying behaviors that are critical to your safety process. Ask, for example, "What might someone do within our proposed safety process if he or she were acting in support of this value?" Then ask, "What would someone do that would undermine the integrity of our safety process?" You should describe interactions that may occur:

- during a normal work day,
- during observations associated with your new safety process,
- during weekly safety meetings,
- during safety committee meetings, and
- during salary and performance reviews.

Before coming together as a group, ask each person to identify approximately three to five practices for each value statement. You may find it useful to have participants record each practice on three-by-five note cards before coming to the meeting. The team then can easily spread the cards on a table and rearrange the practices under the appropriate value statements. You can then allow individuals or small groups to work on combining, clarifying, and refining the practices identified by their associates. When they finish, list all responses on a flip chart, again taping each page on a wall as you fill it out.

Step 2: Sort the practices into appropriate value categories and combine them when possible. After listing all the group's responses, review each list to ensure that each practice is listed under the value that it best supports. Also look for closely related practices that you might combine. The ideal number is between three and seven practices for each value. More than seven makes the list of values and associated practices too long and complex to provide practical guidance to managers and employees.

Step 3: Revise the statements to make them more specific and mutually exclusive. After getting an appropriate number of practices for each value statement, do a final edit to ensure the specificity and clarity of your practices. Try to ensure that your practices are mutually exclusive. If two of your practices are similar or potentially overlapping, either combine them or rewrite them to be distinct.

PHASE 3: PROVIDE TRAINING ON THE NEW VALUES

You will need to provide training to all managers and supervisors, and potentially to all employees, in the practices that define your new values. The best time to provide it is during the implementation of your behavioral safety process. Ideally, it would be the initial module in your implementation training before introducing the safety observation process (see Chapter 10, step 3, on conducting training and kickoff meetings).

The purpose of this training is to clarify the organization's expectations and sharpen employees' understanding of each value statement. You will usually not have to conduct extensive skill development exercises because most employees already have basic interpersonal skills. The purpose of the training is to clarify situations that are appropriate for applying those skills. A typical module might teach how to define behavioral practices, allowing participants to work in small groups to identify practices relevant to each value statement. The participants might then discuss their work in comparison to the practices that your team identified. In some cases, you may need to provide additional training to help employees develop or refine the communication and leadership skills necessary for a successful safety process.

PHASE 4: USE THE VALUES AS BASIC GROUND RULES FOR INTERACTIONS

In your safety process, values have three primary uses:

1. They provide a basis for training employees on how they need interact with one another in the context of the safety process.
2. They provide a basis for safety teams to evaluate the quality of interpersonal interactions in supporting the safety process.
3. They provide a basis for evaluating individual performance within an organization's performance appraisal process.

When your employees complete this initial training, your organization's newly established values should serve as ground rules for their interactions with one another. The statements of values and practices provide a basis for reciprocal feedback and for evaluating actions and potential actions. The question "Did this particular practice support our values?" should be a standard part of evaluating past actions and the question "Does this support our values?" a standard part of decisions regarding future actions. In short, the values can serve as guidelines for the kind of behavior expected within the organization and therefore clarify behavior that members of the organization will support.

Your steering committee can later use these values and practices as a basis for evaluating the quality and integrity of its safety improvement efforts. Chapter 15, on maintaining the behavioral safety process, presented additional details on using your values in the evaluation process.

References

Alvero, A. M., and Austin, J. (in press). The Observer Effect: The Effects of Conducting Behavioral Observations on the Behavior of the Observer. *Journal of Applied Behavior Analysis*.

Alavosius, M. S., and Sulzer-Azaroff, B. (1986). The Effects of Performance Feedback on the Safety of Client Lifting and Transfer. *Journal of Applied Behavior Analysis, 19*, 261–267.

Andrasik, T. (1979). Organizational Behavior Modification in Business Settings: A Methodological and Content Review. *Journal of Organizational Behavior Management, 2*, 85–102.

Andrasik, T. (1991). A Reinforcer Profile in 8 Easy Steps. *Performance Management Magazine, 9*(2), 3–7.

Burns, T. E. (2002). *Serious Incident Prevention* (2nd ed.). Woburn, MA: Butterworth-Heinemann.

Chhokar, S. J., and Wallin, J. A. (1984). Improving Safety through Applied Behavior Analysis. *Journal of Safety Research, 15*(4), 141–151.

Cook, S., and McSween, T. E. (2000). The Role of Supervisors in Behavioral Safety Observations. *Professional Safety*, October, pp. 33–36.

Fellner, D. J., and Sulzer-Azaroff, B. (1984). Increasing Industrial Safety Practices and Conditions through Posted Feedback. *Journal of Safety Research, 15*(1), 7–21.

Fellner, D. J., and Sulzer-Azaroff, B. (1985). Occupational Safety: Assessing the Impact of Adding Assigned or Participative Goalsetting. *Journal of Organizational Behavior Management, 7*, 3–24.

Fox, C. J., and Sulzer-Azaroff, B. (1989). The Effectiveness of Two Different Sources of Feedback on Staff Teaching of Fire Evacuation Skills. *Journal of Organizational Behavior Management, 10*(2), 19–35.

Fox, D. K., Hopkins, B. L., Anger, W. K. (1987). The Long-Term Effects of a Token Economy on Safety Performance in Open-Pit Mining. *Journal of Applied Behavior Analysis, 20*, 215–224.

Geller, E. S. (1984). A Delayed Reward Strategy for Large-Scale Motivation of Safety Belt Use: A Test of Long-Term Impact. *Incident Analysis and Prevention, 16*(5/6), 457–463.

Geller, E. S. (1988). Managing Occupational Health and Safety. Make-A-Difference, Inc., Blacksburg, VA.

Geller, E. S. (1989). Managing Occupational Health and Safety: Marketing and the Human Element. Paper presented at the Fifteenth Annual Convention of the Association for Behavior Analysis, Milwaukee, WI.

Geller, E. S. (1996). The Psychology of Safety. Radnor, PA, Chitron Book Co.

Geller, E. S. (1997). The Hammer Misses the Mark—Why Punishment is Most Often Counterproductive. *Industrial Safety & Hygiene News*, November, pp. 12–13.

Geller, E. S., and Hahn, H. A. (1984). Promoting Safety Belt Use at Industrial Sites: An Effective Program for Blue Collar Employees. *Professional Psychology: Research & Practice, 15*, 553–564.

Geller, E. S., and Lehman, G. R. (1991). The Buckle-up Promise Card: A Versatile Intervention for Large-Scale Behavior Change. *Journal of Applied Behavior Analysis, 24*, 91–94.

Heinrich, H. W. (1959). *Industrial Incident Prevention* (4th ed.). New York: McGraw-Hill.

Hopkins, B. L., Conrad R. J., Dangle, R. F., Fitch, H. G., Smith, M. J., and Anger, W. K. (1986a). Behavioral Technology for Reducing Occupational Exposures to Styrene. *Journal of Applied Behavior Analysis, 19*, 3–11.

Hopkins, B. L., Conrad, R. J., and Smith, M. J. (1986b). Effective and Reliable Behavioral Control Technology. *American Industrial Hygiene Association Journal, 47*, 785–791.

Komaki, J. (1986). Toward Effective Supervision: An Operant Analysis and Comparison of Managers at Work. *Journal of Applied Psychology, 71*(2), 270–279.

Komaki, J. L. (1998). *Leadership from an Operant Perspective.* New York: Routledge.

Komaki, J., Barwick, K. K., and Scott, L. R. (1978). A Behavioral Approach to Occupational Safety: Pinpointing and Reinforcing Safe Performance in a Food Manufacturing Plant. *Journal of Applied Psychology, 63*(4), 434–445.

Komaki, J. L., Collins, R. L., and Penn, P. (1982). The Role of Performance Antecedents and Consequences in Work Motivation. *Journal of Applied Psychology, 67*(3), 334–340.

Komaki, J. L., Desselles, M. L., and Bowman, E. D. (1989). Definitely Not a Breeze: Extending an Operant Model of Effective Supervision to Teams. *Journal of Applied Psychology, 74*(3), 522–529.

Komaki, J., Heinzmann, A. T., and Lawson, L. (1980). Effect of Training and Feedback: Component Analysis of a Behavioral Safety Program. *Journal of Applied Psychology, 65*, 261–270.

Komaki, J. L., Zlotnick, S., and Jensen, M. (1986). Development of an Operant-Based Taxonomy and Observational Index of Supervisory Behavior. *Journal of Applied Psychology, 71*(2), 260–269.

Krause, T. R., Hidley, J. H., and Hodson, S. J. (1996). *The Behavior-Based Safety Process* (2nd ed.). New York: Wiley.

Krause, T. R., Hidley, J. H., and Lareau, W. (1984). Behavioral Science Applied to Industrial Incident Prevention. *Professional Safety, 29*(7), pp. 21–27.

Krause, T. R., Hidley, J. H., and Lareau, W. (1993). Implementing the Behavior-Based Safety Process in a Union Environment: A Natural Fit. *Professional Safety, 38*(6), 26–31.

Krause, T. R., Seymour, K. J., and Sloat, C. M. (1999). Long-Term Evaluation of Behavior-Based Method for Improving Safety Performance: A Meta-Analysis of 73 Interrupted Time-Series Replications. *Safety Science, 32*, 1–18.

Malott, R. W. (1992a). Saving the World with Contingency Diagramming. *ABA Newsletter, 15*(1), 45.

Malott, R. W. (1992b). The Three-Contingency Model of Performance Management. *ABA Newsletter, 15*(2), 6.

Malott, R. W., Malott, M. E., and Trojan, E. A. (2000). *Elementary Principles of Behavior* (4th ed.). Upper Saddle River, NJ: Prentice-Hall.

McSween, T. E. (1990a). Creating a Positive Work Environment. *Chemical Engineering*, June, pp. 135–138.

McSween, T. E. (1990b). Performance Appraisal: What Do We Do Instead? In McLean, G. N., Damme, S. R., and Swanson, R. A. (Eds.), *Performance Appraisal: Perspectives on a Quality Management Approach* (pp. 96–101). Alexandria, VA: American Society for Training and Development.

McSween, T. E. (1993a). Behavior and Safety—The Critical Link. In *Performance Technology 1993 Selected Proceedings of the 31st NSPI Conference* (pp. 191–205). Washington, DC: National Society for Performance and Instruction.

McSween, T. E. (1993b). Improve Your Safety Program with a Behavioral Approach. *Hydrocarbon Processing, 72*(8), 119–128.

McSween, T. E., and Lorber R. L. (1981). Effective Feedback: The Key to Engineering Performance. *Chemical Engineering*, May, pp. 91–96.

McSween, T. E., Myers, W., and Kuchler, T. C. (1990). Getting Buy-in at the Executive Level. In Redmon, W. K., and Dickinson, A. M. (Eds.), *Promoting Excellence through Performance Management* (pp. 207–221). Binghamton, NY: Haworth Press.

Olson, R., and Austin, J. (2002). Behavior Based Safety and Working Alone: The Effects of a Self-Monitoring Package on the Safe Performance of Bus Operators. *Journal of Organizational Behavior Management, 21*(3), 5–43.

Peters, T. J., and Waterman, R. H. (1982). *Search of Excellence: Lessons from America's Best-Run Companies*. New York: Harper & Row.

Peterson, D. (1984). An Experiment in Positive Reinforcement. *Professional Safety, 29*(5), 30–35.

Reber, R. A., and Wallin, J. A. (1984). Validation of Behavioral Measures of Occupational Safety. *Journal of Organizational Behavior Management, 5*(2), 69–77.

Rhoton, W. W. (1980). A Procedure to Improve Compliance with Coal Mine Safety Regulations. *Journal of Organizational Behavior Management, 4*(4), 243–249.

Sherman, S. J. (1980). On the Self-Erasing Nature of Errors of Prediction. *Journal of Personality and Social Psychology, 39*, 211–219.

Sulzer-Azaroff, B. (1982). Behavioral Approaches to Occupational Health and Safety. In Frederiksen, L. W. (Ed.), *Handbook of Organizational Behavior Management* (pp. 505–538). New York: Wiley.

Sulzer-Azaroff, B. (1978). Behavioral Ecology and Incident Prevention. *Journal of Organizational Behavior Management, 2*, 122–130.

Sulzer-Azaroff, B. (1987). The Modification of Occupational Safety Behavior. *The Journal of Occupational Incidents, 9*, 177–197.

Sulzer-Azaroff, B., and Austin, J. A. (2000). Does BBS Work. *Professional Safety*, July, pp. 19–24.

Sulzer-Azaroff, B., and de Santamaria, C. (1980). Industrial Safety Hazard Reduction through Performance Feedback. *Journal of Applied Behavior Analysis, 13*, 287–295.

Sulzer-Azaroff, B., and Fellner, D. (1984). Searching for Performance Targets in the Behavioral Analysis of Occupational Safety and Health: An Assessment Strategy. *Journal of Organizational Behavior Management, 6*(2), 53–65.

Sulzer-Azaroff, B., Harris, T. C., and Blake McCann, K. (1994). Beyond Training: Organizational Performance Management Techniques. *Occupational Medicine, 9*(2), 321–339.

Sulzer-Azaroff, B., Loafman, B., Merante, R. J., and Hlavacek, A. C. (1990). Improving Occupational Safety in a Large Industrial Plant: A Systematic Replication. *Journal of Organizational Behavior Management, 11*(1), 99–120.

Tosti, D. T. (1993). Performance Technology Applied to Organizational Culture and Strategy. Paper presented at the 1993 NSPI Conference, Chicago, IL.

INDEX

ABC analysis, 190. *See also* Antecedents; Behavior; Consequences
  action planning, 198
  identifying weak or missing elements, 202–204
  worksheet, 196–197
Action plans, 201–204
Alevero, Alicia, 18, 216, 223, 237, 240
Antecedent, 190–191, 194, 202
Austin, John, 17, 18, 169, 240
Assessment report, 39–40, 49–50. *See also* Safety assessment
Awards programs, 103–119. *See also* Celebrations; Recognition; Reinforcement
  award programs vs award process and, 106–107
  budget, 116
  cash, 117–118
  contest, 115
  criteria, 108–109
  delivery of, 115
  disadvantages of, 103
  general guidelines on, 119
  identifying potential awards and forms of recognition, 110–115
  internal marketing campaign for, 103–104
  lost-workday injuries, 118
  lotteries, 115, 213
  management and, 116
  OHSA recordable injuries and, 119
  overview of, 104
  paid time off, 118
  presentation to management, 116
  problem with, 11, 12–13, 103
  programs vs. process and, 106–107
  recognition 104
  rules of thumb on, 103
  simple and concurrent, 105, 106
  tiered safety awards, 105, 108–115
  tokens, 115–116, 213
  worker's compensation claims and, 118

Behavior
  definition of, 188
  pinpointing, 188–189
Behavioral safety. *See also* specific topics.
  common configurations for, 90–91
  elements of, 3, 17–18
  implementation stages, 29
  naming, 30, 211
  overview of, 19–20
  support groups, 214
  theory underlying, 188–195
Behavioral Safety NOW conference, 274
Behavior-based safety, see Behavioral safety
Blame, minimizing, 15
Bureaucracy, creation of, 23–24, 143
Burns, T. E., 157

Cards
  feedback support, 212
  promise, 212
  thank-you, 104, 212–213
Case studies
  chemical plant, 255–256
  electric utility, 227–229
  fiberglass-reinforced plastics plant, 259–266
  food processing plant, 237–239
  gas pipeline company, 254–255
  gas pipeline and production company, 223–227
  gas pipeline division, 234–235
  logging company, 229–233
  long term success, 216–222
  paper mill, 256–259
  open pit mining, 266–270
  ore-processing facility, 219–222
  polyolefin plant, 235–237
  self-observation, 223–233
  small company, 234–240
  refinery, 216–219
Cash awards, 117–118

283

Catastrophic events, preventing. See Serious-Incident Prevention
Celebrations, 107, 205–207, 212. See Awards Programs
Checklists, safety observation
 creating, 77–82
 examples, 74–76, 79
 for different areas, 71
 questions that should be answered, 82
 trail-run of, 92
Chemical company, 2–3
 case study of, 255–256
Chemical industry, accident rate in, 5–6
Classroom training, 144
Common problems with safety efforts, 11–13
Compensation, safety incentive, 117–118
Complacency, 1, 8–9
"Condition of employment" policies, 23
Conferences, behavioral safety, 212, 274
Consequences, 191, 203–204. See also Awards programs, Celebrations; Punishment; Recognition; Reinforcement;
 Added consequences, 192
 Built-in consequences, 192
 Factors influencing effectiveness, 194
 Self-provided consequences, 193
Consultants, 35, 37, 274
Contests, 115, 211
Contingency diagrams, 191
Contingency relationships, 190
Contract personnel
 observations and, 88
 orientations for, 210–211
Cooperation, punishment and, 15
Corrective feedback, 16, 88–90
Corrective actions, 161
Crew safety representative. See employee safety representative.
Criticism, impact of, 15
Culture. See Organizational culture

Declarations, public, 212
Delayed outcomes, 193
Deming, Edward, 15
Design process, . See also Design team; and other specific topics
 employee involvement in, 56
 marathon planning meetings 57–58
 milestone schedule, 66–67
 objectives of, 56–57
 periodic planning meetings and, 57–58
 site management team, 59
 steering committee's role, 7
 steps, in, 59
  conducting management review, 125–127
  creating the safety observation process, 68–93
  designing feedback and involvement procedures, 94–101
  developing recognition and celebration plans, 102–119
  establishing mission, values, and mieltsone milestone targets, 61–67
  planning training and kickoff meetings, 120–124
Design team, See also Design process
 implementation plan developed by, 57
 kickoff meetings for, 54
 planning process and, 57–59
 safety assessment and, 39–40
 structure of, 47
 workshop for, 53–54
Drilling company, 2
DuPont company, 4–6

Employees
 accident investigations, involvement in, 209
 addressing concerns of, 140
 design process and, 143
 individual learning history of, 195–196
 involving, 159
 kickoff meetings for, 122–123
 participation of, 18, 69, 240–252
 presentations by, 211
 recognition preferences of, 112–113
 responsibility of, 16
 safety observation process and, 69, 240–252
 safety orientations for, 211
 slogans and logos designed by, 211
 training of, 121–122
 values and, 211
 working alone or in small groups in remote locations, 169–177
 workshops for, 121
Employee safety coordinator, 130
 in self observations, 171–172

False starts, 144
Feedback, 88–90

appreciative, 106
checklist for, 101
corrective, 16
example of, 19
importance of, 17–18
management modeling of 134, 152–153
thank-you cards, 104, 212–213
using graphs to provide, 95–98
verbal, 69
Feedback support cards, 212
Food-processing plant, 237–239

Gas pipeline company, case studies of, 2–3, 223–227, 234–235, 254–255
Geller, E. S., xi, 16, 208
Getting, Jim, 266
Goals, 99–100, 20–204
Graphs
safety observation data, 95–98
posting 95–96, 174
Grindle, Angelica, 219, 234, 235
Group discussions, 211, 213

High-risk areas and activities, identifying, 70–77
Hopkins, Bill, 259, 266

Incident investigation worksheet, 197
Injuries
causes of 6–7
in chemical industry, 5–6
consequences of reporting, 11, 14, 15, 23
investigations of, employee involvement in, 209
lost workday, 118
unsafe acts vs. unsafe conditions as cause of, 6
Implementation process, 29
design teams and, 30
false starts and, 144
implementation plan and, 48–49
management's responsibilities, 134
management support of, 39
objectives of, 56
overview of, 29
planning and, 57–58
process ownership and, 130
schedules 271–273
steps in, 59
team structure for, 30, 39–40, 57
time required for, 58
Improvement projects, 199–205
Incentive programs, 103–119

award programs vs award process and, 106–107
disadvantages of, 103
general guidelines on, 119
internal marketing campaign for, 103–104
overview of, 104
simple and concurrent safety awards, 105, 106
tiered safety awards, 105, 108–115
traditional compensation, 117
Individual learning history, 195–196
Industrial hygiene, 259–265
Involvement procedures, 100–101

Kick-off meetings, 122
Komaki, Judy, 17–18
Krause, Tom, 18

Labels, 189
Layered safety audits, 4–5
Leadership, 145, 159. *See also* Management; Steering Committee.
barriers to, 146–147
effective vs. ineffective, 155–156
informal leaders, 153
leadership checklist, 154–155
monitoring, responsibility for, 149, 151
most important role, 149
new employee orientation, responsibility for, 154
observations and, 152–153
phases, 148–149
positive questions, 150–151
reinforcement and, 154
special role, 147–148
Learning history, individual, 195
Legal issues, 12
Logos, design of, 211
Lost workday injuries
safety awards based on, 118
statistics on, 1
Lotteries, 115, 213

Maintenance of behavior safety process, 136–142
steering committee's responsibilities, 136–138, 178–187
management and, 141–142
Management, 134, 141–142, 145–156. *See also* specific topics
commitment, 159
expectations of, 39

Management (*Continued*)
  implementation responsibilities, 134
  interview of, 42
  involvement in safety process and, 5
  maintenance responsibilities, 141–142
  modeling of observation and feedback practices, 134, 152–153
  monitoring, 149, 151
  observation process and, 152–153
    management review, 126–127
    presentations or question and answer sessions, 214
  process- vs. results-oriented, 22–24
  purpose statement and, 134
  responsibility of, 16, 134, 141–142
  role of, 143
  safety awards and, 104, 116–117
  support of, 39
  training and 52, 123, 124
Management-based safety programs, 13
Management-employee relations
  in process-oriented organizations, 23–24
  punishment and 15–16
  in results-oriented organizations, 22–23
  value-statements and, 27–28
Management overview, 52–53
Measurement and feedback systems 161, See Observation data.
Mission, 24–26
Myers, Wanda, 237

Near miss reporting, 209
New hire orientation, 210

Observation data,
  analysis of, 132, 140, 199
  reviewing, 98
Observation process, 59, 68–93. *See also* Observation data; Safety observations
  advantages of, 69–70
  checklists, 73–82
  common configurations, 90–91
  employee involvement in, 100–101
  example of, 19–20
  feedback as part of, 88–90
  graphing of data, 95–98
  implementation of, 70
    analyzing past incidents and injuries, 70–73
    developing list of critical safe practices, 71–77
    developing observation procedure, 82–88
  drafting and revising checklists, 77–82
  importance of, 17–18, 69–70
  management and, 85, 152
  management review, 93
  management modeling of, 134, 152–153
  starting, 131
  trial-run, 92
  uses of observation data, 132, 140, 199
Observer effect, 240–252
Observer training, 121
Off-the-job safety programs, 214
Organizational culture
  changing, 28
  defined, 21
  model, 21–22

Performance management, 157
Performance standards, 160
Pinney, Ann, 169, 223
Pinpointing, 188
Pipeline company, serious incident prevention case study, 162–168
  causes of pipeline accidents, 164
  critical work practices, 165–166
  reinforcement plan, 167
Problem-solving 199–200
Process, overemphasis on, 23–24
Process owner, 130
Programs vs process, 32
Process safety reviews, 209
Promise cards, 212
Public declarations, 212
Punishment
  appropriate use of, 16–17
  problems with, 13–16

Recognition, 104–106, 205–207. *See also* Awards programs
Reinforcement, 161, 167
Relationships, punishment and 15. *See also* Management-employee relations.
Releases, preventing, 158
Reliability coefficient, 92
Results, overemphasis on, 22–23
Risk management practices, 159
Risks, identifying, 160

Safe behavior promise cards, 212
Safety, as a team process 9–11
Safety awards. See Awards programs
Safety assessment 33–50
  checklist for, 43–45
  definition of, 34

interviews, 42
objectives of, 36–39
outcome of, 39–40
participants in, 35–36
purpose of, 34–35
reporting on, 49–50
safety audits vs., 34
steps in, 40–50
  analysis and development of improvement plan, 48–49
  final report and presentation, 49–50
  observation, 46–47
  review of safety data, 41
Safety conference, 212
Safety index, in self-observations, 172
Safety observations, 68–93. *See also* Observation process
  announcing of, 85–86
  contract personnel and, 88, 210–211
  for employees who work alone or in small groups in remote locations, 169–177
  feedback and, 88–90
  frequency of, 86
  participants in, 83
  procedure, 82–88
  reliability of, 92
  review of data, 98, 199
  self observations, 169–170
  storage of, 88
  timing of, 86–87
  voluntary, 83–85
  who should conduct, 83–84
Safety programs. *See also* Behavioral safety process; *and other specific topics.*
  common problems with 11–12
  components of, 1, 3, 17–18
  management based 13
  management participation in, 49–50, 152–153
  mission statement for, 62
  naming of, 30, 211
  novelty of 32, 107
  off-the-job, 214
  pilot testing, 48, 92
  punishment-based, 13
  traditional, 1
Safety share discussions, 213
Safety slogans, 211
Safety suggestions systems, 210
Safety triangle, 8
Self observations, 169–177
  celebrations, 175

cheating, concerns about, 171
crew safety representative and, 171
safety index, 172
sample checklist, 170, 173
sampling process, 173
posting observation data, 174
prompting, 170, 173–174
recognition, 175
Serious-incident prevention, 157–168
  case study, 162–168
  elements of, 159–162
    building management commitment, 159
    establishing performance standards, 160
    identifiying critical work, 160
    involving employees, 159
    improving the process 162
    maintaining measurement and feedback systems, 161
    reinforcing and implementing corrective actions, 161
    understanding the risks, 160
Spills, preventing, 158
Steering committee, 130–140, 178–187
  addressing process integrity, 181
  behavioral measures, 183
  common problems and, 140
  creating, 178
  implementation responsibilities, 130–139
  improvement targets, 183
  maintenance responsibilities, 136–141
  meetings, 137–138
  member responsibilities, 180–181
  process measures, 179–180
  safety outcome measures, 183,186
  summary of responsibilities, 139, 184–187
  training, 179
Stowe, Judith E., 234, 235
Sulzer-Azaroff, Beth, 17–18, 256
Systematic desensitization, 9

Team structure, 30–31
Training, 65, 121–124
  classroom, 144
  contract personnel, 210–211
  identifying the need for, 123
  introducing new values and, 65
  management and, 124
  new employee orientation, 210
  planning, 123
  steering committee, 124, 129, 179

Thurkow, Tracy 259

Values
    clarifying, 27, 62–66
    defined, 26
    discussions of, 211
    unstructured approach, 275–278

Videos, observer training 121–122

Worker's compensation
    claims, 12
    safety awards based on, 118
Workshops, design team, 53